VIOLENCE AND SPLENDOR

Northwestern University
Studies in Phenomenology
and
Existential Philosophy

VIOLENCE AND SPLENDOR

Alphonso Lingis

Northwestern University Press
Evanston, Illinois

Northwestern University Press
www.nupress.northwestern.edu

Image credits appear on page 161.

Printed in the United States of America

10 9 8 7 6 5 4 3 2 1

Library of Congress Cataloging-in-Publication Data

Lingis, Alphonso, 1933–
Violence and Splendor / Alphonso Lingis.
p. cm. — (Northwestern University studies in phenomenology and existential philosophy)
Includes bibliographical references.
ISBN 978-0-8101-2752-4 (pbk. : alk. paper) — ISBN 978-0-8101-2755-5 (cloth : alk. paper)
1. Phenomenology. 2. Space and time—Philosophy. 3. Art and philosophy. 4. Violence—Philosophy. I. Title. II. Series: Northwestern University studies in phenomenology & existential philosophy.
BD632.L575 2011
191—dc22

2011009137

♾ The paper used in this publication meets the minimum requirements of the American National Standard for Information Sciences—Permanence of Paper for Printed Library Materials, ANSI Z39.48-1992.

Contents

VIOLENCE AND SPLENDOR

Part 1

Spaces Within Spaces

1

Extremes

We let go, let the rhythm take over, and selflessness, voluptuous pleasure, and communion are blended together. It happens in dance and in music. Forty years ago you crossed the Atlantic by ship, and for ten days saw only the ocean surface in moving furrows. You saw and felt all existence in shifting crests and troughs. The oceanic experience is dance and music stripped of everything narrative or anecdotal and become fundamental and immense.

In Bali you got very sick, went to a Balinese doctor, another, a third, and then to a Western doctor; they said it is hepatitis. They said there aren't pills to cure hepatitis. You have to wait it out, weeks, maybe months. You moved into a small hut by the sea. The Balinese are not seagoing people; right offshore the ocean current hurls through the Wallace Trench, seven kilometers deep, patrolled by great sharks. The Balinese live upland and terrace the mountain slopes, irrigating them from the crater lakes of volcanoes. But at the end of the day they descend to the ocean shore, hundreds of them, and seat themselves on the dunes where they wait, silenced by the descending sun. There are always clouds about this mountainous island, and the sun blazes the skies and highlights the cloudbanks with colors found nowhere on terrestrial things. It is each evening different and changing by the minute. You sat among them and the cosmic splendor vaporized your misery and your sense of yourself. You came to realize that there healing was going on and that you were being transformed. Today, thirty years later, you can still remember many of the people and events you encountered those months in Bali, but their memory no longer retains the emotional charge. Yet how vivid still is the cosmic glory of those tropical twilights!

From Tierra del Fuego you took a ship to Antarctica, a big ship because they said a smaller one could not cross these stormiest of seas under winds that gallop around the planet unrestrained by any land barrier. It was twenty years ago. A big ship has a lot of service staff, daily lectures on oceanography, on famous Antarctic explorers, on whales and penguins, and in the evening long meals at big tables where hostesses circulate to keep everybody frenetic with talk. After asking to have your meals taken in your cabin you get such a reputation as a misanthrope that soon nobody talks to you, even greets you anymore. You gaze in silence at the glaciers imperceptibly flowing into the ocean, ice millions of years old, compacted under enormous weight so that the crystalline structure of the ice is changed, and the slabs of light frozen in prismatic colors of violet, lime, azure, indigo glow in your heart like immemorial ecstasy.

Viewed, say, from above, Patagonia is a dagger, with a mountain ridge on the edge of its flat blade, plunging into the Antarctic Ocean. Trekking there, ten years ago, you stomped across stretches of grass so wiry that sheep, they told you, have ground their teeth to the gums on it and have to be slaughtered at two years old. Now, with the market for wool and mutton shrunk, flocks of sheep are few and the homesteads miles apart. Their buildings are made of sheets of corrugated iron, not shaded from the sun, for no tree could stand in the sixty-kilometer-per-hour winds that race across this narrow ridge stuck between one ocean and another. That unceasing wind penetrated you and filled you with its fury and its exhilaration. You could not think of Patagonia without passionately longing to go back, back to that wind.

2

Nature Abhors a Vacuum

The philosopher Immanuel Kant argued that we have an intuition of space—pure space, empty space, extending infinitely—which precedes and makes possible our perception of things extended in space. This intuition is a kind of sight. Kant took space to be something known, all of its properties and shapes and their relations identified by the first complete science, geometry.

But the space about and ahead of us is nowhere empty; it is filled with light, radiant, vibrant, tinted, or somber, or it is filled with the darkness of the night; it is filled with warmth or cold, with dryness or mugginess. When we awaken, in our sleeping bag laid in a meadow, we do not look at the glowing expanse, the warmth about us, the stable ground beneath us. We do not gauge their horizontal and vertical axes or survey their limitless extent. Feeling still opaque and torpid, we are contented with the constancy of the ground that we feel in our body. Feeling the warmth and the radiance of light energizing us, the stability of the ground anchoring and supporting our steps, we stir, we stretch, we get up, we move. We move about because we exist in a luminous and warm medium, supported by the ground, like fish move because they exist in water and birds move because they exist in the open air and the winds.

The light opens before us, marking out the rooms, the corridors; the radiance of the day conducts us outdoors, marking out the paths and the fields, outlining the contours of things. We do not look at the light as we look at things; it orders our movements: our eyes follow the light. The outlying surfaces and contours of the earth summon forth our steps; we advance by following its directions and directives. The sliding waves of the sea direct us to let go of our primate upright posture and slip and slide and flip like dolphins.

Things are not just scattered about in empty space; we find them on the ground, which supports them in their places and supports our positions and our movements toward them. We find things floating in the air, or drifting in the light. As fish swimming about in the lake come upon a tadpole and swallows circling the air come upon a swarm of gnats, we come upon, in the water or in the light, those fish and those swallows and our eyes settle for a while on them, following them in circles of pleasure.

3

Space Travel

Let us stop admiring our heads for a moment, Georges Bataille said, and look our feet. Our big toe is the distinctively human part, the only new body part in the human ape; it stabilizes our upright posture.

Just look at our feet, Bruce Chatwin said; they are long and set parallel; they are made to move on ahead.

At the southern end of Patagonia, constantly battered by the icy winds of the Antarctic Ocean, at a place the Spanish called Punta Gualichó (Devil's Point), we visit a cave on whose walls there are painted dashes and serpentine lines, circles and rows of dots. There are also stenciled outlines of human hands. Some archaeologists have dated them at nine thousand years ago. They can only guess at what the designs might signify. It does seem likely that these broken and serpentine lines and these dots recall the vast distances that the painters had walked to get here.

Thirteen thousand or more years ago, during the last Ice Age, small bands of humans crossed what is now the Bering Strait into the North American continent—or, as now thought, ventured across and then down the coast in small boats. They pushed on, seeing hundreds of miles of glaciers finally give way to the terminal moraines and the green forests and prairies inland. They ventured down the Central American peninsula and lingered at the mouths of rivers that descended the coastal deserts from the Andes; they entered the vast jungles of the Amazon. They pushed on, across so many hundreds of miles of desert and jungle and desert again until here, before the stormy waters of the Antarctic Ocean, they could go no further.

To go as far as you can on our planet, to the opposite flank, you flew from Baltimore to Chicago, changed planes in Seoul, then flew onward to Ulaanbaatar. The thrill of the first men who mounted on horses, the thrill of dashing across vast distances, is anesthetized in passenger jetliners. They are high-tech engineering contrivances to make sense perception illusory: motion of 980 kilometers per hour is felt as fixity in a stationary seat in the felt stillness of the airplane. If you can look outside during daylight, the occasional drifts of clouds below enable you to gauge neither the 980-kilometer-per-hour movement of the plane nor the 1,669-kilometer-per-hour spinning of Earth beneath. Already super-rich tourists are traveling to outer space, to know 190 hours of movement at 39,000 kilometers per hour reduced to an experience of immobility. In a two-week trip by camel into the featureless Sahara extending under blank skies you knew the cadenced pace of the camel over the undulations of the dunes and rhythms in your nervous circuitry and brain that induce the serenity of yogis and mystics. There are no rhythms in the cabin of the transcontinental jetliner and in your body the seat position in which you are buckled builds up muscle tensions that cannot be released. You do the opposite of meditation: your mind sucks on trivial and inconsequential distractions. You watch the in-flight movie or you read a light novel. The 12,250 kilometers to Mongolia are eighteen hours of twitchy urges to do something and a heavy lethargy smothering those urges.

Finally you arrive in Ulaanbaatar.

Although China has annexed the southern half and Russia a big chunk in the north, today's independent Mongolia is still huge, three times the size of France. There are only two million Mongols here, half of whom now live in Ulaanbaatar; the rest are nomads. You asked the only Mongol you knew, the desk clerk at the hotel (what slight knowledge), how to get around and he told you where to find somebody who had a four-wheel drive. When they rent out a vehicle it comes with a driver, though he speaks no English and you of course no Mongolian. But without a Mongolian-speaking companion you could not use the vehicle: once outside the capital there are no paved or even graded roads and of course no road signs.

His four-wheel drive turned out to be a Russian make, twenty-nine years old, and the guy who had the vehicle said the driver knew Mongolia from one end to the other and had friends everywhere. Of course he would say that. And of course if he knows the country and has friends everywhere, he could leave you anywhere with his cronies to cut your throat just to get your camera. There he is, now; you extended your hand to his and entrusted yourself to this stranger who does not understand anything you say.

Trust is taking what is not known as though it were known. (What is currently but loosely called "trust" in the world of governmental, media, and commercial institutions is based on, and increased by, knowledge. Like sampling chemical combinations or physical collisions, a sampling of the operations of institutions in the past is taken as knowledge of their operations in the future. This produces in us, let us say, *confidence* in how the institution works.) There is *trust* where we are aware that the future is not knowable. Trust makes contact with an individual, a power of initiative unto himself but also innately endowed with skills to hide what he does not know, to misinform, and to deceive and dupe others. We trust, have to trust, the surgeon. Before submitting to surgery, we find out all we can about her training, competence, and experience, but in the end we realize that she is capable of misjudgment in diagnosis, neglect of available research, lapse of skill, inattentiveness, or callousness.

Two skydivers leap from an airplane flying 210 kilometers per hour at twelve thousand feet; the one without a parachute falling 150 kilometers per hour has to hold out his arms to slightly slow his free fall, his buddy jumping right after has his parachute and has to dive toward him to hand it to him. Each has full knowledge of what he is doing and has maximum training, but it takes so very little for them to not connect and for the one to plummet to his death—some small nerve firing or some blockage that skewers the convergence of the two men free falling, some slight bristling of jealousy or spite unrecognized in the one with the parachute but that is always there in the closest of friendships and in lovers.

Trust is the strong surge of feeling that connects us with others. Trust is impulsive and immediate. You do not decide to trust someone: you just trust him, or you do not. The leap of trust is exhilaration. There is nothing more exhilarating than trusting a stranger with whom you have no religion, ethnic or moral community, or language in common.

Over the course of the summer you did a circuit of some three thousand kilometers around the country. You roved across utterly flat steppe where never was the least bush to be seen. Seen watching along the horizon on one side or the other were mountain ranges devoid of trees or bushes, and, since it was summer, bared of snow. Mountains that materialize time and patience rather than shape place and landscape. Along them and across the thousands of kilometers of the steppes of central Asia, the winds are on the move, with no end in view. Overhead, the legendary blue skies, most often without a trace of a cloud. When you get up in the morning and look out upon the luminous space without landmarks or obstacles, without promises or threats, you feel such an invitation to move on. You stop for lunch arbitrarily, one spot equivalent to any other. What sights there are are also on the move, though also not going anywhere in particular: herds of horses, yaks, sheep and goats, and, in the Gobi Desert, two-humped Bactrian camels that know no fences or lassos or brandings.

How unlike the vital, essentially practical space that philosopher Martin Heidegger described: here there are no things taking form as the hand reaches out to grasp and manipulate, there are no paths blazed toward objectives, there are not distances made determinate by practical initiatives.

How you understand the nomadism of people here, moving with the herds as they graze across the steppe! How you understand the sudden impulse of men in the springtime to mount their horses and thunder across the open distances, their hearts blazing with sun and wind and speed! In the people you occasionally come upon in this featureless landscape you see an immense, fundamental patience, but as you move on you come to understand they are the most passionate of men. The furies and exultations of thousands of years ago return and press down on them like an immediate and obsessive past.

Those furies and exultations return too upon you. You never just exist in the present. When you awaken in the morning you find yourself in yesterday's landscape, when you move you move in the landscape of hundreds and thousands of years ago. In this now most unpopulated country on the planet, human remains have been found dating back five hundred thousand years. You are joined by the herdsmen who as the last Ice Age was melting down drove their herds of horses and yaks northward and pushed the peoples up north yet more northward and over the Bering Strait to slip down into the American hemisphere; you are joined by those space-intoxicated horsemen who thundered westward to give chase to the Teutonic barbarians who overran and devastated the Roman Empire; by the warrior-nomads who joined the hordes of Chinggis Khan racing over to Korea, across China, down into Java, over to the banks of the Danube, putting all these economies and cultures into contact and circulating their inventions and crafts from one end to the other of the most extensive empire the planet has ever seen, closing off the Silk Road and forcing the silk-coveting and spice-hungry Europeans to their boats to circumnavigate the Cape of Good Hope and overrun India and the Spice Islands and cross the Atlantic to plunder the Aztecs, Maya, and Inca. With these warrior-nomads, in 1347 the Black Death entered Europe, transmitted to humans by bites from fleas carried by marmots, squirrels, and rats, and decimated a third of its population. (This summer too twenty-four people contracted the bubonic plague in the Altaï province while you were there.) These so empty steppes of central Asia are the vortex from which the most momentous events of human history were driven.

In Europe you have trekked Roman roads and in America Inca roads; here the great Khans and their warrior-nomads left no traces of the world-historical events they launched. Here, under the fathomless blue of the sky revered by the Mongols as their only deity, there are no sacred enclaves, no ruins of ancient citadels, no aqueducts, no bridges, no ruined fortresses, no monuments marking great victories or imperial tombs. The burial place of Chinggis Khan himself has never been found, despite teams of geologists combing every square meter of the site recorded in the ancient texts and American and Japanese pilots searching any other possibilities with infrared scans. There are only the tracks of yak and camel herds that the winds will efface a day or so after they and you pass.

The four-wheel lurches down tracks and sloshes through rivers. When it gets stuck, eventually nomads on two-humped Bactrian camels appear and pull it out of a river. At day's end, you are welcomed by nomads into a *ger* ("yurt" is a Turkic and Russian word and not used by the Mongols). A ger is some twenty feet in diameter; the circular wall is made of a wooden lattice frame that can be collapsed together for transport, the roof made of 90 to 110 spokes like a Chinese umbrella; the whole covered of felt made of the white wool of sheep pounded into sheets. It can be taken down and put up in a half hour. The gers are as unmarked as the steppes, the same size, shape, and natural-wool color from one end of the country to the other. But inside you find a little palace: the ground is covered with carpets, the walls lined with embroidered woven hangings in bright reds, cobalt, purple, jade, olive, and yellows, designs traditional in this family and designs marking its history and designs and colors materializing the ingenuity and taste of the weaver. Around the walls there are intricately carved chests for the family possessions. There is a Buddhist shrine veiled with the smoke of incense sticks. You enter into the traditions and memories of clan and family.

There is a low table quickly covered with salted milk tea, breads, cheeses, curds, and pastries for the guest. Soon the ger is crowded with people arriving, relatives and friends, each addressing you with an open smile and ceremonious words of greeting. Chunks of dried dung are lit in the iron stove in the center, pots of rice are cooked, lamb or goat is stewed. You can tell nothing of your wanderings and encounters, but they, regularly refreshed with glasses of *airag* (fermented mare's milk), use the occasion to recount vividly to one another events of their day, tales told by passersby they encountered, recalling tales from long ago. You do not have to know the language to recognize that, seeing a narrator hold the attention as his voice enchains cadences, rises to a thunderous or slows to a whispered climax, seeing the surprise, delight, or satisfaction of a shared memory. Someone begins playing a stringed instrument rather like a mandolin. They will still be drinking airag and telling tales when sleep overtakes you.

By day you come to see the working space of the nomads. A family has horses and yaks, typically two hundred of them, flocks of sheep and goats, Bactrian camels. The horses like the yaks are kept for the milk, mare's milk fermented to make airag, yak milk made into curds and cheese. In the morning the animals are milked and then go off to graze the steppe; in the evening the sheep and goats are brought to the ger camp where prowling wolves can be driven off. The horses selected for riding are fitted with horseshoes, which have to be refitted every few weeks in this rocky terrain. Sheep, goats, and camels are shorn; the wool is carded, woven, pounded into felt; blankets, garments are woven. The goats yield cashmere wool, which is the principal export of Mongolia. Men and boys train in the manly arts of archery, wrestling, and horse racing; there will be competitions on the frequent festive occasions.

This practical layout is set in the social arena you discovered the evening before. A nomad is never alone, even when riding horseback or herding the flocks. The ger is an always-temporary shelter, to be periodically dismantled, but it harbors clan, ancestral, and family traditions, arts, and history. The nomad camp is the arena of circulation of people, where every event becomes a story that surprises and entertains, where places and events from long ago and far away are held in narratives retold and shared. The camp, now here, now there, is a community of humans and also Bactrian camels, horses, sheep, and goats. In exchange for their milk and wool the Mongols devote their labors to them, protecting them from lynx and wolves, rescuing those that have strayed or fallen into gullies, washing and binding their wounds, guiding them to summer pastures in the mountains and to winter pastures in the valleys.

At length you arrive in the region of the highest Altaï mountains in the far West, where the Kazakhs live. They too are nomads, they too live in felt gers, though of different proportions, they too are herdsmen, following their herds and flocks higher into the mountains as the snows melt. Their territory is extended skyward by eagles. Golden eagles, babies, are taken from cliff nests and trained as falcons. Female eagles are kept, for they are the stronger hunters. They have fierce talons; the hunter perches his eagle on his forearm wearing a thick leather glove and sleeve. The eagle weighs some twenty pounds; you have to prop your hand with your other hand to hold her. The eagle hunters are a knighthood; they wear black coats with silver belts and hats made of the tails of fifty mink. When winter comes and the animals are in new thick fur, the hunters mount their horses and climb the mountains and release the eagles, who soar over the mountains and return with foxes, lynx, even wolves. Eagles live thirty or forty years; after ten years the hunters release their eagles into nature so that they can find mates and reproduce.

You come to see that this, the highest country (with the coldest capital), was hoisted up by geologic cataclysms: the great flat steppes visibly were once gigantic lake beds drained as the South Asian continental plate ground its way under the Siberian plate, wrinkling up all these mountains and bursting all these volcanoes. From time to time you come upon deep open gorges where fault lines came apart. You see the sculptures that the winds and glacial melts have carved in these mountains. (How blind that conceit of David Hume's epistemology, arguing that we don't see causality!) One day what from a great distance looked like a thick belt of white fog shimmering under a mountain ridge materialized into a sand dune. It has been measured eight hundred meters high, twelve kilometers wide, and a hundred kilometers long—all in one long singing wall under the black flanks of the mountain ridge. You see the sands drifting onto it; you manage to pull yourself two-thirds up to the top but then feel it give way and you slide with the sands all the way down. You are drifting and sliding in geologic time. In the Gobi Desert you come upon thousands of logs of tropical jungles petrified eons ago. In small museums in the four corners of the country you see some of the fossil dinosaur skeletons found on the sites and some dinosaur eggs; on one fossilized nest the *Oviraptor* parent was crouched over the nest in the position of birds brooding their eggs. The Mongols did not date and explain these features of the landscape as our science teaches us to do, but they observed them with the terms designating causes and events of their account of primeval times. In the rivers your vehicle had to drive into because the river would leave any bridge built aside when the glacial snows melt in the coming spring, and in the unending winds shifting the contours of mountain and steppe, you see the geologic future in the present.

Grasslands, the books called the steppes of central Asia, but wherever you stop, you see a nap of short plants all now flowering in glitzy colors; there are not 10 percent grasses among them. Botanists have identified six hundred plant species in the Mongolian steppes and are still counting. Why so many species? At each stop you find different groupings. There are in fact innumerable biological niches, formed by different soil compositions, different water sources. Your fascinated eyes are held within small ecosystems, spreads of shy, intricately designed tiny flowers; you are enveloped by nature abundant and diverse as in tropical rain forests. You take out a magnifying glass from your pack and between the smallest plants pass into a wonderland of minute lichens and microplants. The diminutive ecosystems are crawling with translucent mites with semispherical eyes honeycombed with lenses channeling sparks of fear and avidity and astonishment. Whole tundras you were about to crush with your foot.

The radius across which you move each day, whose vistas your eyes scan, is one stratum of fractal layouts and ecosystems just under and also just beyond what your naked eyes can see. In them there are no objectives you can want to reach and acquire, no things to detach and refashion with tools and stamp with the spirit, make yours and annex to your identity, will, and status. For you are in Mongolia, in the winds and under the featureless blue sky, twelve thousand kilometers from self-respect and self-importance, summoned to go on and on and on.

4

Learn to Say

Learn to say "Fuck You" to the world once in a while. You have every right to. Just stop thinking, worrying, looking over your shoulder, wondering, doubting, fearing, hurting, hoping for some easy way out, struggling, gasping, confusing, itching, scratching, mumbling, bumbling, grumbling, humbling, stumbling, rumbling, rambling, gambling, tumbling, scumbling, scrambling, hitching, hatching, bitching, moaning, groaning, honing, boning, horse-shitting, hair-splitting, nit-picking, piss-trickling, nose-sticking, ass-gouging, eyeball-poking, finger-pointing, alleyway-sneaking, long waiting, small stepping, evil-eyeing, back-scratching, searching, perching, besmirching, grinding grinding grinding away at yourself. Stop it and just DO.

Don't worry about cool. Make your own uncool. Make your own, your own world.

—Sol LeWitt to Eva Hesse

5

Metaphysical Habitats

A Sea Change in Modern Culture

In 1922 Dr. Hans Prinzhorn published *Bildnerei der Geisteskranken*—translated into English as *Artistry of the Mentally Ill*[1]—in which he reproduced and analyzed 187 from the more than 5,000 paintings, drawings, and carvings he had collected from insane asylums in and around Heidelberg and farther afield, mostly from patients diagnosed as schizophrenic. While psychiatrists before him had occasionally published paintings and drawings of their patients, analyzing them for purposes of psychiatric diagnosis, Prinzhorn found that he could not identify the mental disorders of inmates from the types of images they made. Prinzhorn had earned a doctorate in art history before becoming a psychiatrist; in his study of the works that he had collected he identified six basic drives that give rise to image making: an expressive urge, the urge to play, an ornamental urge, an ordering tendency, a tendency to imitate, and the need for symbols. These, he noted, are the very drives at work in professional artists. Prinzhorn gave predominant place in his book to ten masters, whose works, in expressive power and draftsmanship, he judged to be on the same rank as those whom society recognizes as master artists.

Artists such as expressionists Paul Klee and Alfred Kubin and surrealists André Breton and Max Ernst looked upon the works reproduced in Prinzhorn's book with awe and admiration; Paul Éluard called it "the most beautiful book of images there is." They began at once to incorporate elements of this asylum art into their work. Jean Dubuffet collected a large number of works of the insane and of children, and deliberately set out to paint as the insane do. Wassily Kandinsky, Enrico Baj, and Karel Appel set out to paint like children; Pierre Alechinsky, Keith Haring, and Jean-Michel Basquiat even set out to paint with children. Subsequent contemporary professional art, from the Blue Rider group in Germany to Jackson Pollock, derives more from asylum art than from the Academy.

The discovery of asylum art empowered the search for art among uneducated or naive painters and carvers in Europe, among spiritualists, seers, and mediums, then in prisons (several of Prinzhorn's insane masters had committed crimes),[2] in children's drawings, in primitive or tribal works, African and Oceanic artifacts and ritual objects hitherto exhibited only in ethnographic museums, and in prehistoric and aboriginal rock and cave paintings. Art was recognized in all this "cultural production."

With the discovery of artwork among all these marginal peoples, there arose a new claim to their dignity and a demand for their social recognition. Schizophrenics are not simply deranged individuals who have to be treated by experts or cared for if incurable; they possess positive powers, even creative powers of value to society, perhaps as a result of their disturbed mental condition. "Primitive" or tribal peoples are not simply backward and retarded; their mental and craft productions may well be important and valuable for modern society. Children are not merely potential adults; the child's world has its own structures and its own visions and values that adults may well seek to recover.

The Original Authenticity

The discovery of art in insane asylums in Europe then had a quite different effect from the earlier discovery of art in China and Japan or among the Maya of Central America. What was recognized as art in those far-off lands was, like earlier European art, in the service of religious ideals and had long been crafted according to aesthetic canons. It was produced by trained specialists, on noble materials, and with perfection of finish. Those works could then enter the museums of modern Europe as works of high culture.

The works discovered in insane asylums sparked a radical contestation of high culture. The champions of this art denounced the fine arts academies and the professional artists for producing works that, though they no longer illustrated religious stories and ideals, nonetheless promoted the reigning values and pretensions of the Western ruling elite. Artists, who had lost the patronage of the courts, had become cunning flatterers of the moneyed bourgeoisie. How much flattery went into portraits, into landscapes—how much falseness! In the Academy aspiring artists studied the forms and compositions with which society and nature had been depicted in fine art. But the very recipes derived from the old masters render the works of fine art lifeless. Dubuffet called the works he collected and championed *art brut*—"raw art," contrasted with the cooked, overcooked, cooked-up art of galleries, museums, and intellectuals. His word "brut" also evokes "champagne brut"—Dubuffet had managed a wine business—unsweetened, unsugared art. Authentic art is art not constrained by the canons of the Academy and dictated by the ideas and taste of the cultivated public. Authentic art issues from individuals who are authentic—who are one with their own fundamental and original forces and drives.

If high culture, instead of elevating human life to sublime possibilities, in fact functions to elevate pretense, status, egoism, and greed, the question was whether the primitive state of humankind did not contain resources and energies that high culture weakened and repressed. Energy, feeling, life, creativity would issue from a return to the origins—to primitives, to children, and to the insane, who were seen as still or permanently divested of reason.

Sigmund Freud had compared neurotics with children, and primitives with both, and set out to systematically exhibit repressed infantile drives in neurotic behavior patterns and identify the mental traits of children and neurotics in the mythical omnipotence of thought and belief in magic that are characteristic of primitives. Prinzhorn and the professional artists who acclaimed his book found, however, the fundamental urges that produce art strong in the insane, children, and "primitives," and they declared that these urges produced works of exceptional force because the insane, children, and "primitives" are not constrained by the canons and constraints of high culture and by adult reason.

The artists and critics who championed this work promoted an exalted conception of the artist: artists are creators of new visions, prophets and seers. Their opposites are not the untrained and uncultured, but professionals who work according to received canons of taste and enshrine in their works the received values of society: they are the real philistines. Artists are outsiders, heretics, iconoclasts, revolutionaries. Whereas artists in India, China, and Europe had worked on traditional themes with compositions and styles recognized and transmitted by aesthetic canons, now originality became everything: without originality there is no art. It was to be expected that the more original the work, the more readily it would shock the sensibilities of the general public. Shock value alone—depicting normal values in scatological or perverse guise—does not mean that a work reveals a new vision, but works that are shocking enjoy a favorable first reception among the true artists.

Note that art brut was separated from European folk art, which had also its admirers, historians, and collectors, because folk art is, even more than academic art, governed as to its subject matter, methods of presentation, materials, and styles by tradition. The original wellsprings of creativity would not be found there.

The insane artist, in his or her isolation from culture, is authentic; his or her vision comes from within. The suspension of the constraints of society and reason in him delivers him over to his instincts and visions. The more intense the feelings, the stronger the work. Dubuffet declared: "Madness lightens the man, gives him wings, and promotes clairvoyance."[3] "Insanity is the great art."[4] It was to be expected that originality in the work of these solitary visionaries would be reflected in their behavior—the more original their talent the more original or eccentric their behavior—and it was to be expected that they would be pariahs in society. The champions of their work were disdainful of the psychiatric classifications and stigmatizations of people who exhibit unsocialized forms of behavior; they tended to put psychiatrists and insane asylums in the same camp as the Spanish Inquisition. "We who love them," Paul Éluard wrote, "understand that the insane refuse to be cured. . . . We know well that it is we who are locked up when the asylum door is shut: the prison is outside the asylum, liberty is to be found inside."[5]

What the champions of the art of the insane, children, and "primitives" did retain from Freud was the notion of the unconscious. The creative source of art was now no longer the Muses, sublime or spiritual forces to which the artist ecstatically opens in inspiration; it is instead the unconscious stock of drives and energies with which humans are originally endowed. Art is expressive, expressive of what the artist finds within himself. Or rather, the artist does not "find"—discover and use—these unconscious forces and urges, but releases them with neither constraint nor judgment. But whereas Freud and also Jung conceived the unconscious as impersonal, here the unconscious was seen to be radically individual. It was culture that imposed common traits on individuals—common language, habitual forms of behavior conformed to the standardized tasks that society imposes, and sentiments and emotional reactions conformed to those tasks.

The attention to the art of children was motivated by, and confirmed, the conviction that the fundamental urges identified by Prinzhorn were part of human nature and found in everyone. Ananda Coomeraswamy said, "The artist is not a special kind of man; every man is a special kind of artist."[6]

Prinzhorn, in identifying the six urges in human nature that are at work in image making, had suggested that the image making of the insane gives access to understanding art in general; their work illuminates with exceptional clarity the conditions governing the genesis of artistic activity in humans. The professional artists who incorporated features of art brut into their work or, like Dubuffet, deliberately sought to paint like the insane, sought to put themselves into the mental and environmental conditions in which the insane, prisoners, children, or mediums worked, without becoming insane or being incarcerated. Salvador Dali declared:

"The sole difference between myself and a madman is the fact that I am not mad."[7] The surrealists took from mediums the practice of automatic writing and painted their dreams and nightmares. Paul Klee's work was especially close to the art of children, but he recognized that he worked with dexterity, skills, and experience with media that children do not have. It was through a savant procedure of identifying and eliminating features canonized by cultured taste that he worked.

Distinguishing Characteristics of Art Brut

What was so striking about the works Prinzhorn reproduced in his book that led professional artists to proclaim it "the most beautiful book of images there is"? How did Prinzhorn, who had first been trained as an art historian, identify ten masters from among all the images he had collected? What stood out was the originality and the powerful expressivity of the works. Although they had no art education, these individuals had through long and obsessive application come upon techniques and styles that gave their works both economy and maximum expressiveness. Some of them became extraordinary draftsmen, drawing out complex fractal geometries and gothic architectural constructions with surety.

Although it was the news that art, real art, pure art was to be found in insane asylums that was responsible for the far-reaching effect of his book, Prinzhorn did not title his book *Artistry,* but *Bildnerei—Image Making—of the Mentally Ill.* These insane image makers, for the most part schizophrenics, were oblivious of the whole system of academies, galleries, museums, collectors, critics, and intellectuals that separates art from mere craft or simple making. The staff of the asylums in which they were incarcerated viewed the drawings they made, often on paper bags or toilet paper, and the figures they shaped out of chewed bread as of no interest and cleared them out and destroyed them at the end of the day. Most of the works of those Prinzhorn identified as his ten master image makers had been destroyed. And indeed the makers of these images did not make them as art. Nor do prisoners, spiritualists, children—nor indeed do tribal peoples, whose productions were classified in ethnological museums as ritual objects, ceremonial paraphernalia, magical implements. John M. MacGregor explains:

> It is of crucial importance to understand that the motivation underlying the production of these powerful objects or images almost invariably has *nothing* to do with "art." These "things" are the product of an ob-

> sessional involvement with images in the service of extremely unusual preoccupations and ideas. Let me give some examples:
>
> - Maps, of planets or worlds they have visited, and which they now wish to present in diagrams or images;
> - Religious icons, but of private religions having little to do with any religion we may be familiar with;
> - Embodiments of repressed but explosive sexuality, made for sexual reasons, not as art;
> - Cosmological diagrams, illustrations of theories of the origin of this, or another alternate universe;
> - Medical illustrations of bizarre biological and anatomical concerns;
> - Influencing machines, or perpetual motion machines, depicted as plans (or actually built), or magical weapons for defense against imagined enemies;
> - Spirit depictions of the "other side," inspired visions sent by the departed, or by God. . . .
>
> Commonly, there is no desire to share these things with anyone, and the images are kept secret, or are hidden. This is an art called into being in a desperate attempt to map or to depict unexplored, and occasionally frighteningly unfamiliar, aspects of human reality. Frequently, examination of the entire body of work of a single artist reveals an ongoing involvement with the creation of an alternate reality, another universe, conceived in encyclopedic detail.[8]

When, in 1995, a museum of art brut was opened in Baltimore, it was consequently named American Visionary Art Museum.

Although critics have not been able to identify distinctive features—of subject matter, materials, forms, colors, techniques, or style—of art brut as opposed to the art of the art world, several traits are quite common. There is a predominance of figures in two dimensions, and an absence of the representation of depth through the use of perspective—figures are not graded in scale to represent their positions in depth. There is also a tendency to completely fill up all the space. The image is saturated; however terrifying the figures, the terror and the anxiety are all there, completely laid out. Frequently an inmate spends years on the same theme, filling it out more and more completely. The aim is not "art," is not perfecting the depiction of an object, but materializing a personal environment, where the prosaic common environment of ordinary people is blocked out.

Critics have also noted that most often faces are presented in frontal view, their sides symmetrical, features inexpressive, stark and stereotyped, and eyes open wide as though staring at the viewer. This is also common in the carved figures, statues, and masks of African or Oceanic tribal art. In everyday life, facial expressions give us clues about the thoughts or mood of the one we encounter. In these works the face is staring directly into our eyes, revealing nothing of what he or she sees, yet holding us from surveying our environment and him or her on our own, making us aware of another environment that is holding these eyes transfixed. Roger Cardinal suggests that we think of the everyday experience of bumping into someone in a doorway. Abruptly halted, we gaze a little longer than decorum permits into the eyes of a stranger. At this moment the environment in which we were making our way is suspended, and we become aware that the other sees something different, which we become aware of only through an emotional shock.[9] The frontal staring faces that these works so often depict ensnare our gaze and turn us to a habitat unknown by us and alien to us.

Georges Bataille opposed forces in us concerned with survival and socialization, utilitarian drives issuing out of our needs and wants, with excess forces produced by the strong and healthy organism that have to be discharged, discharged without recompense. Artists were seen as devoting themselves to discharging their creative forces in works without utilitarian function, gratuitous, ends in themselves. This would be true only in the measure that art quite recently was disengaged from religious, political, social, and ethnic causes and enterprises. The champions of art brut exposed as hypocritical the professionals who produce high art: they channel the fundamental urges that give rise to making images into making images for others; they put these urges at the service of profit, prestige, celebrity. The insane, prisoners, and mediums and spiritualists who make art brut have, out of desperation, put those six fundamental urges that produce images at the service of visualizing, materializing a metaphysical habitat for themselves.

Recognizing, and Endangering, Authenticity

Prinzhorn wished to protect the image making of the asylum from being dismissed as the delirium of the mad by demonstrating that the fundamental urges that give rise to their image making are identical to those that drive professional artists. The urgency of this was evident from the start; as soon as the works of the insane were published and declared to be artworks by avant-garde professional artists, the defenders of the Academy mocked those works as delirious expressions of deranged individuals. The time was advancing rapidly when their work will be put in Nazi exhibitions of degenerate art and publicly burnt. Most of the asylum masters Prinzhorn had identified disappeared during this time and were presumably included in the Nazi program to liquidate the incurably insane.

What the champions of art brut discovered in asylums was authenticity, not only conceived negatively as works free from the pretense, greed, and ambition that falsify the professional art of high culture but also conceived positively as powerfully expressive of individual obsessions and visions, surging up from the positive forces of a distinctively individual unconscious.

It was at once realized that recognizing these works as art put this authenticity in danger. They would enter exhibitions, galleries, museums; amateurs would collect them; they would acquire monetary value and their makers fame; their makers would enter into the social stratum of cultural producers. All this could only put their authenticity in danger. Prinzhorn, who identified ten master image makers in the asylums in and around Heidelberg, did not reveal their names and attached pseudonyms of his contrivance to their works. Dubuffet set out to acquire thousands of works from authentics and eventually housed them in an institution in Lausanne, Switzerland, set up as an antimuseum. They would be thus kept outside of commerce. And outside of the art world; they would not be lent for exhibitions in art museums elsewhere. Alain Bourbonnais purchased hundreds of art brut sculptures from galleries and exhibitions and put them in farm buildings he had acquired far from any cultural centers.

Prinzhorn had concealed the names and biographies of the master image makers. But if these works were seen to emerge from the individual forces of the unconscious and owe their exceptionally strong expressive quality to the force of those compulsions, separating them from the whole of the maker's biography could only be artificial; these same forces must have also shaped the maker's conceptions, emotions, and actions, releasing them without the constraints of social conformisms. Works that were done for or in sexual excitement surely could not be separated from the maker's biography, since so much of society's mental and physical energies have gone into regulating the drives of sexuality. Adolf Wölfli, a schizophrenic semiliterate peasant who has been recognized as among the greatest of the asylum artists, had been incarcerated as criminally insane after repeated sexual assaults on very young children.

Dubuffet recognized the urgency of protecting the works of art brut from the art market. But declaring them to be artworks, as did professional artists who incorporated features of asylum art into their work, was to bring the art world to them. Professional artists like Max Ernst and Arnulf Rainer set out to acquire their personal collections of these works, as Picasso and Matisse had collected African and Oceanic masks and fetishes. If the works of Klee and Kandinsky were to acquire ever-higher prices in the art market, should not the works that inspired them also acquire such monetary value?

Before long the names of the masters in the asylum were revealed. Before long it was acknowledged that the recognition awarded to them outside should not be hidden from them, and that they should receive their just share of the big money publishers, collectors, and galleries were garnering from their works. As this happened, the makers began to react to the appreciation of their works, to the selections of collectors, and began to draw, paint, and carve in response to them. Dubuffet began to separate the genuinely authentic works from these subsequent works adulterated by the tastes of the cultured.

It is an obvious task, not yet undertaken, to systematically itemize the nature of these adulterations—to identify what conscious calculation does to the unconscious drives seeking to materialize metaphysical habitats, identify just what culture does to the six basic drives that give rise to image making. It is indeed surprising that professors of aesthetics, which from its beginning separated fine art from craft, decoration, mere illustration, and the designing of commercial lures, have not been busy at this task.

Disintegration of the Ideology Involved in the Recognition of Art Brut

The classical Freudian equation of neurotics with children and with primitives came into disrepute. Primitive art had been seen as primitive in that it had not gone through the European Renaissance discovery and incorporation of perspective and the succession of materials and technologies—fresco, ceramic glazing, oil painting on canvas—and the codification of the canons of taste that had evolved European high art, as tribal people themselves were primitive in that they had not gone through a history of sedentary and urban culture, nation building, and industrialization. But field anthropologists came to recognize that every culture dubbed "primitive" has in fact undergone a history of technological and cultural adaptation to its environment, and could have survived as it did in often inhospitable conditions only if it had made such successive adaptations. Hunter-gatherer tribes in the Amazon were found to be survivors of sedentary and urban centers destroyed by the Spanish conquistadors. Instead of sharing with asylum art the character of being the direct expression of the unconscious drives of individuals, the masks and ritual objects prized and collected by artists in Europe such as Picasso, Braque, and Breton embody the taste, ideals, and cosmology of a particular society. Field anthropologists found that these are works of skilled and recognized specialists in those cultures; they could not be taken as representative of the sort of creative impulse that Dubuffet held was a universal human endowment. They were anonymous only because the collectors had deigned acquaintance with the cultures. Once it was recognized that the art brut of the insane, the imprisoned, and spiritualists and mediums are in fact maps of planets or worlds, cosmological diagrams, icons of private religions, medical illustrations, influencing machines, magical weapons, or depictions of visions and of visitations of the dead, these works cannot be put together with the drawings and paintings of children.

The champions of art brut had hailed the source of art in the unconscious, a wellspring of creative energies found in everyone, even in children and the insane—found especially unadulterated in children and the insane. But already Prinzhorn recognized that asylum art is rare. He had selected to locate ten masters out of the five thousand works he had collected. Already these had been the result of a selection; he and his associate Dr. Karl Wilmanns had written psychiatrists and asylum directors that he was interested only in works that "are not simply copies of existing images or memories of their days of health, but intended as expressions of their personal experience," and are "outstanding individual achievements."[10] He recognized that many, even most, inmates simply show no urge to draw or paint.

What is more, the sources of asylum art seem to have dried up. With all the interest concentrated on art brut, master image makers are no longer discovered in asylums. It is therapy that is blamed. For the delirious forces that had expressed themselves in the intensity of powerful works are nowise relieved or cathected by these works; to the contrary, these works visualize and materialize for the maker the alien and abysmal elements that he had divined in his environment and intensify their fearfulness and enigma. They prolong and deepen his delirium. Now the generalized use of antipsychotic and mood-altering drugs reduces the intensity of emotions and drives. The widespread introduction of art therapy does not produce art brut; to the contrary. True art brut was a self-taught and in most cases secret activity, where the use of materials and the strokes were stabilized and perfected through obsessive repetition. Enjoining art activity from the outside, even though the art therapist shows no models and encourages the patient to draw or paint however he or she wants, inevitably results in works without inner compulsion or skill. It is enjoined in order to aid the patient to recognize his disordered perceptions and feelings; the recognition will issue in judgments and constraints. What this produces, John M. MacGregor reports, "is simply amateur art; mediocre, cliché-ridden, and dull."[11]

Something similar has been found for children's art: works of young children that are strong and expressive are in fact rare. And no sooner do children acquire dexterity and some experience with the materials, the mass of public images crowds in on them to guide their drawing and they too produce works that are mediocre, cliché-ridden, and dull.

Dubuffet had found art brut also outside the asylums—among the uneducated, in rural areas, among people who had no real contact with the canons and commerce of the art world and who had not been apprenticed in folk art traditions. There the individuals who devote themselves to image making have for the most part begun to make images late in life, sometimes after some trauma, or simply after retirement. But today when television and advertising have invaded the most remote or backward corners of the land, it is less and less likely that individuals, even isolated in rural areas, can be found who sketch out drawings that are not guided by public and commercial models.

Outsider Art and the Eclipse of Authenticity

The great exhibition that opened in London in 1979 was titled "Outsiders," no doubt because there is no English translation that carries the connotations of the French "art brut." It did restrict itself to the works of people isolated from the professional art world either through incarceration or socioeconomic marginalization and untrained in the methods and canons of high art. Prinzhorn had originally separated the works from the biographies of the makers, to the point of concealing their names, but because art brut could not, like Impressionism or Cubism, be identified by common internal characteristics, it was biography that alone identified a maker as an outsider and therefore the work as outsider art. In fact the term "outsiders" was becoming a broader and also more questionable category. Indeed the great London exhibition brought these works inside and to the very center of the art world.

Resorting to the criterion of works produced outside the institutions and commerce of the art world—academies, museums, galleries, auctions, art historians and art critics—inevitably deflected attention to different regions of lowbrow or popular culture: postcard photographs, New Age images, videos, computer draftsmanship, jazz and tango, pop music and its album-cover and poster imagery. This production had been disdained as popular art and commercial art, and grudgingly still called art only by virtue of the quality of craft that went into it. It was taken to respond, not to the refined taste of the cultured but to crude emotions and tastes. But once the irrational drives and emotions of the insane had been recognized as the source of creativity in art brut, the crude, raw emotions and elementary notions with which the masses appreciate popular and commercial art could be seen to have kinship with them.

Avant-garde professional artists such as Andy Warhol and Roy Lichtenstein turned to this production with an appreciative eye. As earlier artists of the avant-garde incorporated features of art brut into their works, and even set out to paint like the insane, now the avant-garde selects images, colors, media, and techniques out of popular and commercial image production to renew and expand the domain of art beyond sculpture and painting in fresco, on canvas, and on paper. Popular art is fashioned to provoke a response from large numbers of people; commercial art uses colors, movements, and sounds to fix logos and advertising slogans obsessively in memory. Professional artists manipulate popular and commercial images to neutralize their commercial effects and provoke simple amusement, produce ironic or sarcastic effects, or convey images and messages of social or political revolt. Thus the avant-garde artist, however celebrated, however rich, maintains his status as an outsider—heretic, revolutionary.

British artists Jake and Dinos Chapman do not see any essential distinction between popular and commercial art and the art of today's avant-garde. Commercial art functions less to exhibit the real use value of commodities than to cast glamour or prestige over commodities of banal, dubious, or bogus utility. It is less the real shampoo or vacation package we buy than the image created by that art. "Modern art is the absolute elaboration of capital," the Chapmans say, "because it's an object that can be bought for a huge amount of money without any intrinsic value. . . . So someone like Jay can say, 'Right, this is £1 million or £10m,' whatever, and once someone writes the cheque it has acquired that worth. If people are gripped by a romantic discourse about the meaning of a work of art, then all we can say is that the sublime is very expensive. You should pay well for it."[12]

Another Obvious Task for Professors of Aesthetics

One day in a shopping mall I noticed that a big discount store was going out of business; everything was on sale for half price. Thinking there are always things one could use in the house—towels, new kitchen chairs, dishes—I strolled about for an hour. I left without buying anything—driven off by the idea of bringing into my home any of these towels covered with pastel flowers, these pink plastic kitchen chairs, these dishes overdecorated in gaudy or insipid colors. I left thinking if, as aesthetics now tells us, art is all around us, in pop and commercial image and spectacle production, why is the greatest part of what is made mediocre, cliché-ridden, and dull? On the way back to my home, the question stuck: Why do they make all these suburban houses looking mediocre, cliché-ridden, and dull? Why have the rich bought these oversize barnlike mansions with their fake-Tudor details, set in manicured lawns with neat foundation plantings?

One obvious answer is that the mass marketing of consumer goods requires designing them for the lowest common denominator. But that only raises the question: why do the masses, and the rich, want consumer goods that are mediocre, cliché-ridden, and dull?

The evident direction to turn is to sociology: since the masses acquire a sense of their identity and belonging by acquiring collections of consumer goods recognizable by others who appreciate such collections, conformity to the fashion and taste of a group is paramount. The rich seek out rare and one-of-a-kind things, Ferraris and yachts, designer clothing, original works by celebrated artists. They even acquire revolutionary, contestatory works that reveal new visions and consequently new ways of life. They triumph over them by appropriating them, by exhibiting with a smile a piece of graffiti art in their elegant drawing rooms.

But the sociological explanation only raises the question that we could call psychological, or the question about human nature. The champions of art brut were convinced that innate to every human is a creative urge, seated in the unconscious wellsprings of life. These are the innate strengths that make each human an individual. But there are also deep-seated weaknesses: are not people who fill their home environments with overstuffed, overdecorated furniture, Disneyland ceramics, and racks of golden oldies CDs driven by nostalgia for more familiar, more comfortable times past? The idle rich surround themselves with antique furniture and artworks that appropriate the exceptional labor and talent of rare individuals. Bad taste is produced by character weaknesses.

Then the artistic impotence of "normal" people is unhealthy, betrays deep-seated dysfunction, and reveals the institutions and procedures that make them normal, adult, rational as pathological.

Persistence of the Longing for Authenticity

"Authenticity" is a term that has disappeared from the vocabulary of our postmodern philosophers and from our cultural studies, occupied in explaining the cultural constitution of race, gender, person, Western individualism, and art brut.

Do we not have the everyday experience that others looking at us, talking about us or to us, are addressing the identity and worth that we have acquired with our accoutrements: they see and address the American, the professor, the inner-city street kid, the suburban SUV owner, the fashionable socialite. When we go to far-off places inhabited by members of our biological species, are not what our eyes see of them but tourist images—facial contours, complexions, and garb that look exotic only to us? And when someone there is standing before us, speaking directly to us, we have been cautioned that he is not speaking with his own voice but speaking the language of his gender, his family, his class, his education, his culture, his economic and political interests, his unconscious drives, indeed his state of physical health and alertness. The effort to know him gets detoured into efforts, ever more evidently fragmentary and superficial, to know all these layers. Today the professionals who study these things write books exposing how superficial and deluded have been the efforts of the experts: exposing the imperialist, Christian, Victorian, Romantic, or Orientalist fables written by those people who left their homelands and fell in love in some remote place, married, and never returned; the positivist, Freudian, or Marxist fables of the last generation of cultural anthropologists; the rationalist, structuralist, or postmodern fables of the current generation.

But when we confront the works of art brut, we feel, we see something else: we see stark stereotyped faces with eyes staring at us making the environment we survey fade away and indicating another world; we see the visualization, the materialization of a world unlike anything we have ever seen; we feel the vehemence of anxieties, terrors, explosive sexuality; we read words, the overfamiliar words of advertising slogans or biblical injunctions, here meaning something strange. However much these works have been museumified, interpreted by academics skilled in explaining the cultural constitution of individuality and authenticity, these works are disturbing. Inwardly we feel the mysterious forces of suffering, of yearning, of desperation. They are not the solutions invented in desperate situations; they are expressions, visualizations, intensifications, materializations of desperation.

And the concept of authenticity returns to haunt us.

The images made by the insane, obsessives, and mediums are not artworks, made for the public, for us. They are works so often made in secret. These works force us to understand the radical difference between what they, or any artwork, can mean for the maker and what they can mean for the public, for us. Already Jean Dubuffet realized that to call them art brut—art—is to resituate them among us, where for us they acquire a different function, significance, and value. Yet can we not join in what these works really mean for their makers? Do we not to some measure always project ourselves into the strange and alien habitat that the maker has visualized and materialized? Do they not draw us into the dark cauldron of his desperation?

Permanently on display in the Baltimore American Visionary Art Museum is a gigantic canvas eight feet high by seventeen feet wide in seven panels depicting two Michaelangelesque lovers themselves made of thousands of human figures painted with intricate and ever-varying detail and confident skill. The label tells that this is the only work of James Franklin Snodgrass who created it over decades and left it to his cousin when he died. It was by chance that a researcher came upon it and eventually acquired it for the museum. Each time I go there I sit to contemplate its radiant colors and fractal composition, then block off another square foot or so of it to examine close up, as the painting extends indefinitely beyond me and ahead of me. I see that I am caught in it and that this will go on for years.

One year the museum exhibited a room of pictures all 2.25 by 2.75 inches, enclosed in thick frames, pictures so small that one had to view them with the magnifying glass provided. Their maker, Ray Materson, as a youth had been arrested for cocaine and sentenced to twenty-five years in prison.

Doing time—it means doing nothing, enduring time. His mind had no future with which to occupy itself; it returned to the life he had lived, his childhood. Having to periodically repair the prison garb he was forced to wear, he found himself repairing also the garb of other inmates. He remembered his grandmother, who sewed. Who occupied herself the long winters indoors with embroidery. He began embroidering pictures of prison life, using for colors the unraveled fine threads from discarded nylon socks; the minuscule pictures he embroidered occupied very long stretches of time.

Peering with magnifying glass in hand, we rejoin him with the inmates behind the prison walls. How can we not, contemplating these scenes walled in with thick frames, made of millions of painstaking needle strokes, rejoin him, the future and the outside world so wiped off his life, in his desperate endurance of time, his finally transfigured suffering?

Encountering these works we do not simply appreciate them formally, aesthetically, but experience them as excitants to infantile, criminal, sexually explosive, and visionary compulsions in us. They materialize the zones outside of the common world that excite such compulsions, the abodes that harbor them.

6

Departures

In the nineteenth century the structural malady diagnosed in individuals and also in society was conceptualized with the term "alienation." Dysfunctional and sociopathic individuals were called alienated, conceived to be possessed by alien forces, though these aliens were no longer identified with supernatural beings and the alienated no longer treated with exorcism but with psychiatry. The malfunctioning of modern society was analyzed as alienation: in capitalism workers are dispossessed of the products of their labor and dispossessed of their desires, understanding, and will; it is those of the capitalist that act in them. Twentieth-century philosophers found alienation in the Socratic, metaphysical, and Christian doctrines that our body and this world are not the true home for humans.

The critique of alienation engendered reflection on what it would mean, positively, to not be alienated. The concept of home became the central concept in Martin Heidegger's work. Heidegger declared that for humans to exist is to exist in the world, and explained that we are not simply in the world like a rock is in a meadow; we inhabit the world. We inhabit the world by making a home for ourselves here, by building. In Heidegger building and dwelling become the very definition of human existence.[1]

Building for Heidegger is assembling, assembling earth and skies, mortals and immortals. While assembling stones and boards and thatching or tar roofing we assemble the supporting and sustaining substance of the earth. A house is open to the skies, assembling on itself light and darkness, warmth and cold, rain and sun. A building is a shelter for us because we are mortal; it harbors our vulnerability. It assembles what is good and wholesome, and brings near the giving that lays out the world and that is holy. We exist in the world by building, assembling an environment, a world.

For Heidegger dwelling with beings that we do not construct is also assembling. Farming is building up one's herds and flocks and building up the fertility of the land.

Nature is the array of natural resources with which we build; it is also, Heidegger says, the sun whose position our initiatives take account of; it is the wind "in the sails"; it is the inclement weather we situate in time and place with our covered bus stops and railway platforms.[2]

For Heidegger all the fundamental activities of our existence are aspects of building a home here. Everything we do is for the future; we build homes for our children and grandchildren; we build for history. We build for immortals, to receive among us the harbingers of the healing and the wholesome. Heidegger's conception of building and dwelling is also a conception of thought and truth. Thought, which in the metaphysics established by Socrates served to liberate us from our bodies and this world and put us in touch with the realm of the ideal forms, has instead to be seen to serve building a home for ourselves on this earth and under these skies. The supporting and sustaining earth, the illuminating and vitalizing skies, the vulnerability and mortality of humans, and the harbingers of immortal things are brought to light through building and dwelling. Truth is set forth not only in discourse but also in institutions and buildings.

Yet ours is not a sedentary anatomy; our feet are made to move on. Heidegger does not recognize our compulsion to leave the home that we had built for ourselves, to leave the landscape in which vital human interests and spiritual values have been embedded, to depart for realms unmarked by human intentions and manipulations. We go off to the nearby or far-off forests, the mountains, the glaciers, the beaches, the oceans. We make our way across mountains and continents as the continental plates collide and buckle up those mountains that freeze the west winds and dry out these deserts. We descend into the Grand Canyon treading the eons that deposited those fifty layers of petrified sediment. In the crystal nights of deserts and mountains our gaze travels the light-years of the stars. We visit excavations and monitor the millions of years from algae to dinosaurs. We trip through the savannah with the wildebeests and the impalas and stroll the beach tacking the waves with the plovers. Spaces, just because they are exterior, exotic, enthrall us and summon us.

Having to find food, seek shelter, perceive and foresee dangers, and take means to protect ourselves, humans develop an essentially sound knowledge of their vital environment, its resources and possibilities, develop a memory of the past and a reliable anticipation of the future and knowledge of causalities at work. This is the sphere of empirical and pragmatic reason. It must be free of fantasy and credulity in essential matters. But we also depart, in rituals, visions, and trances for a cosmic realm with a different kind of reality.

Anthropologist Clifford Geertz sees that religion, myths and ritual, do not arise out of anxieties in the economic and political sphere; they are not simply magical supplements to instrumental methods of production and methods to enhance and enforce political systems.[3] What drives the religious departure for a cosmic order are, he says, the chronic failure of our explanations in the face of the ravages of nature such as volcanic eruptions and plagues, the portentous visions of dreams, and puzzlement about the origin and place of humans in the world. But also innumerable everyday perplexities make us chronically uneasy. Physical pain, personal loss, worldly defeat, or the helpless contemplation of people's agony drives us to envision a cosmic order where suffering is explainable and thus becomes something that has to be and can be endured. We strive, have to strive to work out some normative guides to govern our actions, but see all too often that ethically correct behavior results in disaster, while behavior we can nowise approve of is rewarded. The religious perspective envisions a wider, cosmic history that accounts for the fallen or corrupt nature of our world, which so often thwarts our efforts to live according to sound moral judgments.[4]

The world in which we build our habitat is the world in which we are most solidly rooted, whose inherent actuality we can hardly question, and whose requirements we can least escape.[5] The cosmic order envisioned in religions is real in a different sense and in a different way from the common sense world; it does not simply extend further the common sense world. One *departs* for the supernatural world.

Entry into and perception of the reality of the cosmic order are produced by ritual. Rituals make people experience the presence of extraordinary forces; in rituals the powers represented in the myths are experienced, Geertz says, as *presences*.[6] Rituals engender adherence not first to the propositions of a creed, but, in tribal religions, to the persuasive power of traditional imagery; in mystical ones to the apodictic force of supersensible experience; in charismatic ones to the hypnotic attraction of an extraordinary personality.[7]

Rituals engender extraordinary moods—the solemnity and grandeur of Monlam, the Great Prayer Festival, at the Barkhor in Lhasa in Tibet; the delicate eroticism of the Gerewol of the Wodaabe in the Sahel. Moods are not hazy sensations within the closed sphere of our minds; they have a cognitive function. They reveal how we are embedded in the reality of our environment and disclose how the environment as a whole affects us, weighs on us. The environment is dark, heavy, oppressive, or light, buoyant, exhilarating, or bleak, undifferentiated, desolate, or cascading in brutal surprises. The rituals articulate and give expression to moods—to grief and mourning, to triumph, to expectation and hope.

And rituals articulate longings into motivations for certain kinds of action. Plains Indian rituals that stage flamboyant courage set forth motivations to fast in the wilderness, to conduct solitary raids on enemy camps, and to thrill to the thought of battle. Melanesian Manus rituals that dramatize moral circumspection implant in participants tendencies to honor onerous promises, to confess secret sins, and to feel guilty when vague accusations are being formulated in the community. Javanese meditation séances induce inclinations to maintain one's poise in awkward or frustrating situations and to experience distaste before emotional outbursts.[8]

Motivations, emphasized and fixed in rituals, direct us to specific layouts of goals and means. An action posits belief in the resources and motivations for that action. We see the world as troubled by capricious and malignant foes, or we feel the reality of compassion in specific events and in the general benevolence of Nature.

Interned in insane asylums, incarcerated in prisons, confined to the margins of the social world where the economy circulates and education advances, as the makers of art brut work, often for years, on the same subject they withdraw more and more from the reality of common sense, of other people and their doctors and caregivers, to dwell in the metaphysical habitat they visualize and materialize for themselves. This habitat elicits their most vehement passions, and it also confirms and intensifies these terrors, ecstasies, loves, and lusts.

When we recognize the authenticity and sincerity of the works of art brut, and the authenticity of their makers, we will say that, unlike the celebrity professional artists who do not believe in the paintings they make to flatter their patrons and flatter the tastes of their clients, the makers of art brut believe in themselves, believe in their fervent passions, and believe in the metaphysical habitat they visualize and materialize for themselves.[9] The beliefs of these persons are not formulated in statements, and after the images have been made they typically do not formulate in words what they represent. Their belief is intrinsic to, realized in, their gestures, their making of the images, in something of the way religious belief is formed in performing ritual. The moods that are identified and expressed and the motivations that are formulated in religious ritual rebound into the practical common sense world, but here through the making of images these individuals withdraw ever more definitively from that world. But while the makers of art brut do not return to the world of our common sense, we discover that we can find our way to their metaphysical habitats.

7

Plumage and Talismans

Coiffure, cosmetics, scarification, and tattooing make the outer surfaces of our bodies into artworks. Perfumes, the softened or stentorian intonations of the voice, the melodic lines of gestures, and dance extend body artistry across space and time. We do not hesitate to exclaim of a fish, a pheasant, or a horse how beautiful he or she is, but "How beautiful I am!" is something we must only whisper and when we are alone. We are concerned with our body's appearance, but it is bad taste to show it; our body's beauty should be ingenuous, harmonious, and unselfconscious, like the beauty of plants, butterflies, and birds, like the beauty of the dawn and of twilight.

Things shape our bodies, imprint their forms on them. Raiment, headdresses, and jewelry make our bodies elegant or seductive, gallant or authoritative. We put the forms of our bodies on things. We make clay, marble, paint, chemical emulsions into statues, portraits, photographs. These things that we have shaped shape us in turn. Some of them hold smooth or grainy or patchy images of our bodies that are all surface and color, divested of depth. They make us move as patterns in the world pageantry of colors and forms. Some of these things may materialize anatomical ideals or dignity, authority, the radiance of deities. Their forms take possession of us and choreograph our movements.

In found objects come upon by chance, by luck, or made, that bear some resemblance to our bodies or that do not—unusual bones, fossils, crystals, burls of trees, carvings—we feel their inner space and the vibrancy in that inner space, strong and healing. We wear these things on our bodies and they make contact with the inner space of our bodies.

In the cave paintings of Lascaux, Cosquer, and Chauvet, where lions, rhinoceroses, aurochs, horses, and bears in movement were depicted with consummate skill, the few depictions of humans are stick figures shown dead or, when alive, with the head of a bird or antelope. These depict not the outer appearance of our bodies but a connection with the forces of the inner space of bird or antelope. The earliest known sculptures of our bodies, figures such as the "Venus" of Willendorf dating from the Upper Paleolithic, are concerned not with showing how our bodies look from the outside but how they feel from within. The first, Gandhara, Buddha images were made in Afghanistan under the influence of Hellenic statues, which had long depicted the appearance of particular individuals, but the Buddha images were never meant to depict the outer aspect of the man Siddhārtha Gautama; they depict an inner life in composure and compassion, which they induce in the viewer. In Africa and Polynesia masks did not only change the outer aspect of the body but were experienced to intensify the density and turmoil of its inner space. In Bali and Tibet, donning the masks induces trance.

Overleaf: *Man Asleep*

8

Inner Space

In 1981 sculptor Antony Gormley exhibited, in the Whitechapel Gallery in London, *Bed,* twin beds made of 8,640 slices of Mother's Pride bread, with impressions of his body as if lying in each bed.[1] The slices of bread Gormley had removed to make these impressions, he ate. Eating had emptied the body from the bed—or had made visible on it the inner space of the body.

For the next twenty-five years, Gormley had plaster casts made of his body. He held a position that was inactive, standing hands at his sides and feet together, or crouched in a fetal position, or lying on the floor. These casts of his body would then not exhibit dignity and authority like statues since the time of the Egyptian pharaohs, or make narrative gestures and show heroic deeds like the statues that populate our public buildings and parks. Holding himself immobile while the plaster hardened, Gormley experienced the inner space of his body, a zone of energies, tensions, and intensities. "Action can be confused with life. Much of human life is hidden. Sculpture, in stillness, can transmit what may not be seen."[2]

When the plaster cast was cut open and he stepped out, Gormley soldered thick lead sheets around the hollow left by his body. The facial features were smoothed and rendered anonymous. The hands and feet were also smoothed over in the lead sheets, taking from the hands their agility and propensity to grasp, immobilizing the feet. Lines of soldering crossed the body cases like lines of longitude and latitude of planets adrift in outer space. Sealed in these body cases, the figures appear solitary and self-contained. Yet their inactive, passive positions show their vulnerability. Behind the thick body cases, we sense their nakedness. The bluish-gray lead both reflects and absorbs light, giving a sensitive and sensuous character to the figures. And the lead marks them with a specific vulnerability. Lead is a metal used to block radiation, as we who have been administered X-rays in dentists' offices know. But lead is poisonous. These lead body cases, enclosing the spaces left that a body had occupied, invoke Gormley's and the public's anxiety over the thermonuclear arms race and the imminent possibility of the extinction of human life.

***

Working with an inanimate, inert substance, Gormley discovers the distinctiveness of the space of our bodies. When we awaken and open our eyes, space extends about our body. In the light and across the ground we see planes, fields, and horizons, and an array of things within reach and paths open to further things. The things about us are situated along axes of close–far, up–down, in front of–behind, and right–left. These axes extend from our body. The things about us can be approached and viewed from different positions; as we turn about them, their sides anticipate one another and enclose those things. We measure where they are with our hands and legs: they are within immediate reach or reachable by short or long, easy or difficult movements.

We perceive our body to be in the same space as the things about it, but not as an extended and closed shape alongside of their shapes. We can observe only parts of our body, not its overall shape. In its image in a mirror we can see more of its surfaces and shape, but the image is at a distance and we still perceive our body occupying the space where we are, in front of that mirror, as a shadowy and elusive substance.

Although we cannot perceive the bones, muscles, and organs behind our skin, we do have an inner sense of our bodies. When we stand up and take a position, when we walk or climb, and when we manipulate something, we have a kinesthetic sense of the postural axis of our body. This kinesthetic sense maintains our weight in balance and orients our force and momentum in a direction and toward an objective. Through this sense of the postural diagram, we have a sense of how our feet are spread as we stand on an unstable ladder or shove our weight against a car that is stuck in the snow.

About this postural diagram we have a quasivisual perception of the volume of our body. As we walk we quasi-see our body in the forest path; when seated, our legs stretched under the table have a kind of virtual visibility. We look at the narrow bend in the cave and quasi-see our body able or not able to pass through it. Psychologists have called this our "body image." It is not the same as the image of our body that we could see in a mirror, nor is it the image we could project by a mental act. As a postural axis forms within our body, it generates this virtual visibility about itself. As our postural diagram tightens and extends with eagerness, hardens with rage, loosens up in contentment, becomes willowy in sensuous languor, our body image dilates or contracts, becomes more dense or more supple.

When we stop orienting ourselves toward objectives and operating implements, we can let our gaze drift in the array of things about us. We can also turn our attention inward, to the inner space of our bodies. We cannot view its contents from different angles and positions, but we feel the intensive forces of our life pulsating in it. In feeling these intensive forces and in feeling their attachment to this space, we have a fundamental sense of being an individual.

Antony Gormley remembered that as a child lying awake in bed he had long savored the inner space of his body. He recalls a time when bicycling that he felt that he was rising and his body extending in height. During three years spent in India, Gormley, learning *vipassana* meditation, practiced attentive absorption in this inner space. He found that in letting go of the concerns and tasks of the environment, the sense of a distinct ego as a pole of purposive initiatives, an ordering and organizing force, fades out, while the sense of life, his life, becomes more intense and vibrant.

In his London studio Gormley attended only to holding his body immobile as the plaster hardened—an experience repeated in hundreds of hours spent in making hundreds of plaster casts of his body. Certainly the experience of being sealed off from the outside environment by the plaster hardening on the naked surfaces of one's body can seal one in anxiety.[3] But Gormley found that because he had done this voluntarily and knew his wife was there to cut him out of the cast when it had completely hardened, he experienced the inner space of his body as a zone exultant with intensive forces. This inner space felt, he says, like a seed or bomb.[4] It "implants itself within our consciousness and in some way [is] the reason for an explosion of new sensation or feeling of aliveness."[5] Each time when the plaster cast was cut and he stepped out of it there was an experience of being born.

Beginning in 1999, in a series of works titled *Quantum Cloud,* Gormley surrounded the space where his body was with a chaos of steel tubes, like trajectories of energy. In the empty core space we see the body materializing and disintegrating as a condensation of these throbbings of energy, or see them emanating from it. "While keeping the body still and non-gestural, it is active, energetic, mobile."[6] A thirty-meter-high *Quantum Cloud* that Gormley made in 2000 was set in the River Thames.

It was essential for Gormley to cast his own body for the figures. The inner space of our bodies is not something that our minds grasp conceptually and it is not something whose shape and properties we determine from a conceptual, cultural, or perceptual context. It is only accessible through direct physical relationship with it. The anonymity of the body cases refers us to the inner space that they contain and that is singular, known and felt by itself. Here is an individuality quite different from that which we find in the particular cast of the face and the complexion and in the postures, gestures, and deeds we see on the external appearance of a body. Gormley is concerned not simply with representing it in a form accessible to vision, but with experiencing it and inducing a like experience in us.[7]

When we stop tracking a specific gesture or operation that another's body is making and contemplate his or her body, we can sense something of its inner space filled with life. We can even get a sense of the inner space of life in contemplating inert things—such as these sculptures. The immobility of the material of the sculptures and the immobility of the positions of the figures holds us. We can hardly look long at Gormley's sculptures without sensing what it is like to be sealed in these body cases, held in these positions. "I want people to feel they're inside the work," he said.[8]

The inner space of our bodies does open upon the space of the environment. *Three Ways* (1981) showed bodies curled into a ball, bending over, and lying down with holes at the mouth, anus, and penis—"the primary portals from the body to the world." In *Land, Sea and Air II* (1982), the figures are standing, kneeling, and crouching with eyes, nose, and ears open.

Sites

The making of statues has long entailed making plinths, pedestals, niches, and shrines to support and concentrate the forces of the statues and building chapels and temples to extend their forces sovereignly across the environment. In our times the blank walls of museums and the white cube of galleries were taken to best liberate the powers of a statue.

When Gormley put his sculptures, anonymous, immobile, exhibiting neither initiatives nor gestures, in galleries and exhibition halls, they were as out of place there as our bodies are in the white cubes of the high-rise apartments of our megacities. A figure seated with trunk at a right angle to the joined legs is set with its back on the floor; another is bent in seated position on the floor but on its side.[9] A figure is standing horizontal on the wall;[10] another clings to the wall like a fly.[11] In a disaffected prison in Charleston, South Carolina, five figures have risen so that their heads have penetrated through the ceiling.[12] In Malmö, Sweden, the inner spaces of bodies, not cast from Gormley's body this time but from those of three hundred people of all ages, are enclosed in blocks of concrete and set in rows as in a military cemetery.[13]

In San Casciano in Tuscany, in a cylindrical water tower a hundred meters high and ten meters in diameter, Gormley set an iron cast made from his standing body some ten meters from the ground, horizontal against the inner wall of the tower. But here the figure is not struggling with its confinement by the circular wall of the tower. As soon as we

Lost Horizon I

enter it we throw our head back and our upward look pivots us in this horizontal position of ascent. Stairs spiral up the inner wall, across small windows, to the top where from the sky the tower platform shows the tapestried panorama of the Tuscan landscape.

Most of Gormley's later works were installed outside of galleries and exhibition halls, in natural environments or in urban settings. Outside, these figures no longer appropriate and subordinate or struggle against the space in which they are put.

We have long been familiar with the proud statues that mark landscapes, mountains, and cliffs along the oceans with the imprint of human presence and that mark cities and lands to belong to conquistador and king and their god. But there are so many sculptures that instead of fixing the human presence in the outlying spaces reveal what is unappropriated and alien in those spaces. Such are sculptures in sacred, sacrificial places and in burial places. Such also are sculptures showing lookout points and entries into mountain ranges, deserts, and oceans; such are the lines of flags strung across mountain peaks in the Himalayas, showing the winds and the heavens.

Another Place consists of a hundred sculptures of iron that had filled seventeen molds of Gormley's body held in an immobile position and differing only slightly, with individual traits of his features and his body effaced. They were first set up in 1997 over an area of 2.5 square kilometers in mud flats outside Cuxhaven, Germany, which was one of the major ports for emigration to America in the middle of the last century.[14] The figures look outward toward the horizon. Daily the tide submerges them; then they emerge again. Seaweed grows about them and mussels attach themselves to them. Visitors are invited to wander from one to another and to look with them toward the horizon of better places. We, with the sculptures, stand on this shore and are submerged and emerge from the mud of our life.

Inside Australia has fifty-one sculptures Gormley set up in 2003 before and in Lake Ballard, a salt lake near Menzies in western Australia. The figures look out upon the shimmering heat layers off the salt flat left when the waters of the lake dried up eons ago. They lead us visitors into this geologic time.

For the figures of this work Gormley enlisted fifty-one volunteers, both Aboriginal and immigrant, who live in the environs. This time the molds of their naked bodies were not made by coating them with plaster but by scanning them with an electronic scanner. Then a computer reduced the thickness of the head, trunk, and limbs by a third; these were then cast in an iron alloy. Gormley calls these figures "Insiders." "The Insiders are the body reduced to one third of its mass that acts rather like dark matter, matter subject to excessive gravitational force, that becomes like an absolute antithesis to the surrounding space," he explains. "An Insider . . . is a core rather than a skeleton. It is a way of allowing things that are internal to the body—attitudes and emotions embedded in posture or hidden by gesture—to become revealed. They are equally alien and intimate."[15] Set in the Australian desert before the ancient salt lake, they show the way the body space condenses in order to maintain itself in the limitless expanses of the desert about it.

In Poggibonsi, Italy, Gormley made plaster casts of the bodies of six townspeople and one visitor (*Making Space Taking Place,* 2004). For the molds these bodies stood still, hands at their sides, not holding a particular attitude or gesture. Gormley's assistants then set, in random arrangements, blocks of four sizes, each eight times the volume of the previous one, within the confines of the molds, and then cast these sets of blocks in iron. The figures show the inner space of their bodies without the outer surfaces that confined it. In these inner spaces of bodies we see not a functional organization, but closed units, pixels, in random order. As we look at them, the assemblage bounces our eyes from one closed unit to another, producing a sense of vibrant potentiality.

The blocks in this inner space of a body echo the paving blocks and bricks of the buildings about it; indeed these sculptures look like relief maps set vertical of the town. The rough-surfaced iron of which they are made makes them appear to rise out of the molten core of Earth.[16] Their pixilated forms refer us to the digital computer processes with which things and buildings in the outlying space will be designed and constructed henceforth. The sculptures are diagnostic instruments set up to reveal the city and the landscape to which its inhabitants belong. They are guideposts leading us into deeper layers of the geography of the town, marking intersections of ancient pilgrimage and trade routes that gave rise to the town, and to its repeated destruction and reconstruction. They mark sites where the emotions of its inhabitants make contact with this deep structure.

The sculptures make us see that the bricks and blocks of the town, the Earth beneath us, and the digital technology ahead of us have an intimate connection with the inner space of our bodies. The sculptures make us wonder what this connection is and what it will be.

Inner Communication

Attentiveness to our outer appearance issues in communication: I who see but patches of the front of my body see the bodies of others as integral wholes and can observe them from all sides, and I see that others see me in that way too. Gormley discovers that the inner spaces of our bodies also communicate. "Part of me would like to believe in the Teravada Buddhist tradition—that you can transmit love as a vibration which is independent of any object. It radiates out. It is registered in that space of the darkness of the body, rather than in the other world, the world of daily life, external appearances."[17] In works such as *Present Time* (1987–88) and *Bearing II* (1995), the lead body cases are fitted into one another such that the inner spaces of bodies merge.

Field (1991) inaugurated a series of works[18] that exhibit the communication of the inner spaces of bodies and that were themselves collaborative works. *Field* consisted of thirty-five thousand hand-size terracotta figures made according to Gormley's directions by the Texca family, brick makers by trade, in Cholula, Mexico. The makers soon found they were making figures each in his and her individual way. The figures were to be made quickly, which contributed to give them an individual stamp. "The extraordinary thing was that everyone found their own way of doing that and also how the clay held the feeling and took on personal mannerisms," Gormley remarked.[19] *Asian Field* (2003), exhibited in diverse locations in China, contained 190,000 sculptured figurines made by 350 people of all ages from the Hudau District in Guangzhou. "It is made of a hundred and twenty-five tons of clay, energised by fire, sensitised by touch and made conscious by being given eyes," Gormley says.[20]

The figures without arms, legs, and mouths are mute and defenseless. They are reduced to the inner space of their bodies. They are small, but the space of their bodies and their sheer numbers fill up the exhibition halls. Their eyes are turned to the viewer, eyes that appeal and question us. We look at them, look into their eyes, and our eyes take up their appeal and their question, a question about the sense of life in space, its vulnerability and its enigma. This art provides, Gormley says, "transitional objects to help people bridge the change from a social matrix based on labor and hardship that forged the strong sense of community to one based on pleasure and the shared experience of potential and possibility."[21]

Part 2

Snares for the Eyes

9

The Fallen Giant

The sign does not say when the sequoia fell. Or why. Perhaps it died of old age. The sign says that the oldest giant sequoia found is 3,267 years old. It says that giant sequoias (*Sequoiadendron giganteum*) are the largest single organisms by volume on the planet. The sequoia named General Sherman stands 274 feet high, its trunk is 102.6 feet in circumference at the base, and it weighs some 6,100 tons. The sign says that this tree stood about 220 feet high. Its trunk at its widest is 72.6 feet around. You stand with raised hands next to it and then walk the length of it and back to get a sense of its length and girth. You feel the great volume of the fallen tree next to the inner space to which your life is attached.

Blue whales grow to 110 feet long, weighing 220 tons. You had seen, in Patagonia, the fossilized skeletal form of the biggest dinosaur yet discovered, *Argentinosaurus huinculensis,* 130 feet long, weighing some 100 tons. This sequoia grew to 220 feet and weighed 4,500 tons.

The forest trees you are familiar with back home seemed to you to rise high in order to be able to spread their leaves in the sun. In open fields you saw their life in the thousands of leafy and flowering branches; their trunks were the support of those splayed branches. Here you see the life of the sequoias about you in their massive trunks plunging upward; you see the branches, many of which die as the trunk continues its millennial ascent, and the small scaly leaves existing for the sake of the trunks. You wonder over this thrust in life to ascend. You feel it in your body that rises from the ground to stand upright and looks skyward.

The life attached to the enormous inner space of the sequoias, to these hundreds of tons of matter, is somehow one. One life governs the system you see in the branches, a lateral branch subordinate to the upward drive and coordinated with the other laterals on its level. The ordered spacing of the branches is modified, here on this level, there on this side, by the space just here occupied by branches of other trees and by the mix just here of sun, shade, and fog. A branch does not go off on its own, but each branch has to adjust to local conditions and events. The mighty trunk itself, while maintaining its cylindrical form slightly narrowing as it ascends, at each stage has had to adjust to the impact and pressures of events. Swerves of bark mark these adjustments.

When this tree died, it died everywhere, in its massive trunk, its branches, its millions of rootlets. Not one of these branches inventing on its own its detour around obstacles or its advance into a clearing of sunlight had a life that could continue on its own.

The sense of life attached to the enormous inner space of a sequoia, or to that of a beached blue whale, dominates our perception of their surface forms and colors. Our sense of our life attached to the inner space of our bodies communicates with the inner space of these enormous beings. Our wonder over the tiny volume in which the intense life of guppies or sandflies is attached overwhelms our fascination with their external designs and colors. The lizard *Sphaerodactylus ariasae* is but 16 milimeters long. *Wolffia globosa* and *augusta,* but 0.6 milimeters long, produce exquisite flowers and are very prolific; one plant could produce 1 nonillion (1 with thirty zeros) new plants in four months, a volume of flowering plants equal to the size of Earth.

The term "substance" in ancient philosophy—something that stands in itself, subsists of itself—functioned as a very general term to designate the beings external to us that we take to be real. In modern times philosophers pronounced us incapable of knowing the substance, the nature, or the essence of things; what we perceive of them, they said, are only sensuous patterns. We know not things themselves, but appearances. Appearances do not subsist in themselves; they exist in relation with our sensory surfaces and organs. Empiricist psychologists had supposed that these appearances can be decomposed into discrete sense data produced by the impact of stimuli on our sensory receptors. But phenomenological psychologists observed that every perception is the perception of a pattern against a background. Our sensory receptors form one interconnected system that focuses our sense organs and sensory surfaces on a pattern. The patterns shift as we move. Patterns that change their shape and hue without changing their apparent location as we move about them appear as sides or "profiles" of things. What we perceive as "things" are coherent and consistent sequences of patterns. The phenomenological account distinguishes between patterns, sides or "profiles," and "things" that are sequences of surface appearances. It reduces our experience of reality to a field of surface patterns that appears to be laid out at a distance from our eyes.

The distinction between appearances and things that appear is peculiar to vision and does not really have analogues in the realm of sound, taste, odor, and the tangible.[1] In *Being and Time* Martin Heidegger argued that it is through manipulation, using, that things take form for us. Perception is not a simple recording of patterns, but a survey of paths and obstacles as we move and a focusing on objectives and implements. We see the forces and resistances of things in their colors and textures: the sidewalk looks firm and hard; the muddy ground looks soft and sticky. Our environment, he said, is a layout of forces, whose reality is known by and is reducible to the effects they have on other forces.

The materiality of things, in Heidegger's interpretation, is in their supporting or resisting our manipulations. Their forms are the ways their forces fit into other things and into our manipulating hands. Their efficacity is in the way our forces find that they gear into the forces of other things. The hammer's nature is revealed by hammering; what is discovered is not the hammer's "properties," that is, traits that belong to it, but its "appropriatenesses," its specific ways of fitting in with our hand and of driving certain kinds of nails into certain kinds of wood—and its recalcitrances. The finality of things is their susceptibility to sustain the movements we make to acquire further things. The human agent himself is not to be seen as a material substance that periodically launches movements; he ex-ists into the outlying world, exists as a succession of purposive manipulations of outlying implements, a succession of circumspective, goal-oriented surveyings of the layout about him.

Beyond the layout of paths, implements, obstacles, and objectives in which we are actively engaged lies the region of what is on hand (*Vorhanden*), available, at our disposal. It consists, Heidegger explained, of what has been left aside and what has broken down, been used up, is in the way.

Yet is not this a strangely narrow picture of our experience? In every glance about us, our eyes, our ears, our touch, and our sensibility record patterns, rhythms, tonalities, reverberations, mists, glows, glimmers, sparkles that we are nowise manipulating or using. When we look at the butterflies, trees, and mountains in their independence of and indifference to us, we see them as they are.[2]

We do experience the field about us as a field of forces, but not first and only as vectors of force, instrumental relations. Space is filled with the radiant or waning light and the brisk or humid air. As we move among the branches of the trees and the bricks of a building we sense the intensive forces with which they occupy their volumes. When we come upon a giraffe in the African savannah, the phenomenological account would detail the comprehensive survey that records its surface colors in their contrast with colors in the surrounding field, and Heidegger's account would describe its shape as a dynamic force in movement toward objectives ahead of it. But we have an intense sense of the inner space of life in that giraffe.

Antony Gormley's sculptures induce in us a sense of the inner space of our bodies. With a heightened sense of the inner space of our bodies we communicate with the inner space of his lead or iron-alloy sculptures. We also communicate with the inner space of other inanimate things.[3] When the heat of high noon in Giza drives the tourists into their air-conditioned buses taking them to air-conditioned restaurants, we slip back into the Great Pyramid of Khufu and find ourselves alone in the tomb room of the pharaoh. We have an intense sense of the small expanse in which our life is attached, and of the enormous expanse of rock of the pyramid around us, under us, over us. We are safe; this structure has stood for forty-five hundred years. But the incalculable disproportion between our inner space and the space of the pyramid compresses us.

We perceive the inner spaces of things when we tentatively step across the fissured ice slabs of the glacier. When we watch the river plunge over the cliff. When we sit on a boulder in the Arctic tundra and rest our eyes on the boulders about us. When we study the design of the snowflake holding a droplet of water.

10

The Stone

On my last day in the Highlands of West Papua, a Lani man I came to know gave me his precious possession, a black stone some four inches long strangely marked with thick white lines and thin white lines intersecting in the zones the thick lines squared off. It is oval in shape and smooth and nestles in the palm of my hand. He gave me to understand that he carried it with him whenever he left his compound to go into the jungle, and was giving it now to me to guide me on my onward wanderings. He had seen that every day I wandered about without any discernible objective or goal, and understood that I had wandered over the planet to New Guinea without any task or project.

Older white travelers would say that this stone is for the Papuan a fetish. What is it for me? It is not really a cherished souvenir of his friendship; in fact I can no longer remember very clearly what he looked like. Yet I am so often drawn again to this stone; it summons me.

11

The Voices of Things

We find ourselves summoned forth by the earth, the light, water, and fire; we find ourselves directed by the things about us. Animism and fetishism designate two different ways we have understood these experiences. Animism—from *animus* (wind, spirit, voice)—recognizes a spirit in material things. This spirit is separate from them. The Book of Genesis opens with the Spirit speaking in the void and ordering the waters above and the waters below, seas and dry land, flying and swimming, creeping and crawling things. Yahweh speaks to Moses from the burning bush. In Greece the oracle speaks in the vapors rising from the cleft rock at Delphi. Since the spirit, or spirits, can speak in things anywhere, animism sees our whole environment, living and nonliving, to be a realm of meaning.

The form given to the voice can convey information about things near and far. A voice can also address, call upon, and order us. Our linguistics recognizes the vocative and imperative force of speech acts; our grammar and rhetoric classify their forms and uses. But our sciences, even our psychology, do not account for the fact that a voice uttered at a distance can penetrate to the core of our identity and appeal to us and put demands on us. A child may be ordered by the voice of a parent when he or she is no longer there; an adult may hear that voice when the parent is no longer alive, may hear too the parents of that parent. Indeed, is there any society where the voices of ancestors are utterly silenced? Animism recognizes that voices addressing us and ordering us may be the voices of other species, and voices of the absent and the dead. They can be heard in things.

Fetishism recognizes a silent voice of material things themselves.[1] Things lure us, provoke us, direct us, charm or hex us. The voice that is heard is only in this singular material thing, which we come upon by chance. Fetishism recognizes a realm of good and bad luck.

We find ourselves in a partly or largely manmade environment whose structures, tasks, and paths were planned, and we design our actions and follow maps and signs. Yet even there, we encounter nourishing, energizing, and enchanting things and sinister and baleful things by chance. Strokes of good or bad luck, they lead us into byways and freeways from which we may not return to our planned objectives. We plan our day, program our ambitions, diagram our life, but, looking back, we may recognize that the essential turns our life took were determined by chance encounters—a captivating teacher, an opportunity that opened and that brought us to a work we love, someone met by chance who became our lover and life partner, a debilitating accident, a disease. That we were born with our sound or sickly body, with our particular sensibility, that we were born at all were strokes of luck.

Images of the Fetishist World

In the cave paintings of Chauvet, dated 32,000 B.C.E., Cosquer, 27,000 B.C.E., Altamira, 18,500 B.C.E., and Lascaux, 15,000 B.C.E., the art has already reached mastery in the expressive use of the supporting rock, the shadings of pigments, the composition of forms, the use of split perspective to put the figures in movement. Pablo Picasso, in viewing these techniques at Lascaux, said, "We have learned nothing in twelve thousand years." Aurochs, bison, tigers, antelope, horses are portrayed in movement and depicted with astonishing anatomical skill. Yet typically there are no images of humans. In Lascaux there is a stick figure of a man with the head of a bird. In Chauvet there are none; in Cosquer there is one stick figure of a dead man.

Humans did leave the mark of their hands. In these caves and caves in Patagonia, in the Sahara and the Kalahari, in Indonesia, and in Australia, peoples with no cultural links covered rock walls with stencils of their hands, made by holding their hands to the cave wall and blowing dry or liquid pigment across them. Handprints made by simply dipping one's hand in colored clay and pressing it against the cave wall are rare. Archaeologists are baffled both by the universality of the hand silhouettes found in caves continents apart and by their possible significance. Should we see in them the impulse to mark inert substance with the mark of human presence and will? But the stencils outline the absence of a hand. The hands vanished, leaving only the rock, or perhaps they vanished into the rock.

There are peoples who never marked and designed the landscape in which they lived. The Maasai of Africa refuse to wound the earth; they do not chop into the ground to plant gardens or fields but instead live off the milk of their cattle and wild plants they can gather, drawing the blood of bulls—in small quantities so as not to weaken them—to supplement their diet. They do not cut into the earth to make bricks, and instead shape their dwellings of dung from their cattle.

The Aboriginal Australians find that the contours and hollows of the continent have a melodic form, and capture it in their songs. Having learned the melodies from their parents, they find their way from one end of the continent to the other by following its songlines. They do not mark the land, but find the landscape already marked with traces of events of the Dreamtime. The monolith Uluru, a ruby-red rock 3.6 kilometers long and 348 meters high, is found planted in flat desert that extends for hundreds of kilometers on all sides, as the very heart of the continent. In immemorial repose and silence the rock sleeps—and dreams. It dreams of pythons and scorpions, blue-tongued lizards and wallabies. The Australians have over generations elaborated a dream analysis of its forms, its furrows and its hollows. There are paintings in many of these sites, some made on the ground and obliterated by dances performed on them, some in caves made forty-two thousand years ago. The paintings bring out, and care for, the dreams of the sites themselves.

In India the core of the world was seen in concentrated and pure form in the Shiva linga. It is an oval black stone, with three white lines near the top, set in the yoni, a "womb" not into which the "phallus" plunges but out of which it rises. It is associated with Shiva, who dances in a ring of fire the unending dance of creation and destruction. The most sacred Shiva lingas, few in number and celebrated throughout India, are those made in the heavenly Ganga, that luminous cosmic torrent (which Europeans call the Milky Way) that Hindus see descending to Earth at Pashupatinath in the Himalayas, where the terrestrial Ganga, the Ganges River, begins and then travels across the subcontinent to plunge into the ocean and reemerge in the most remote heavens. The most sacred celestial Shiva lingas are found at Pashupatinath. (These celestial Shiva lingas Europeans call meteorites.) The second most sacred Shiva lingas are terrestrial, formed by the Ganges River; they too are few in number. In a cave at Amarnath in Kashmir is found a Shiva linga of ice that reaches its fullest each year in July and summons tens of thousands of Hindus from all over the subcontinent who come to contemplate it the night of the full moon. Finally there are the millions of Shiva lingas carved by human hand. On the small island of Gharapuri (Elephanta) off Mumbai there is a cave whose walls have been abundantly carved. On the day of the equinox, the setting sun sends a beam of light that glows on the Shiva linga in the core of the cave. The cosmic and the terrestrial Shiva lingas are primary: in the Hindu conception they could not be images of human genitals "projected" onto them; instead human genitals are tertiary images of them. Rather than marking the human presence in the river plains, the caves, the mountains, and the sky, the Shiva linga marks the human anatomy with the Ganga, the cosmic river.

The Buddhist stupa Borobudur in Java, the greatest monument erected in the southern hemisphere, encloses a small mountain. Located in jungle, you have to take several local buses from Yogyakarta to get there. The time it takes to visit it makes clear what it is doing, and what it is doing to you. As with every stupa, you circumnavigate it clockwise, but here the path rises spiraling each time higher in ten levels as you pass the walls covered with hundreds of bas-reliefs depicting all the activities of humans. As you rise higher and higher upon the stupa, the activities depicted become less agitated and more harmonious. You reach the top; there are no more bas-reliefs and you find yourself encircled by seventy-two bell-shaped stupas, whose walls are perforated to show inside figures of the Buddha, but the stupa that is on top at the center is empty, showing nothing, or everything, inside. You look down upon the luxuriant jungle below extending on all sides as far as your eyes can see. To the south the jungle ascends a wall of volcanoes, simmering fumes and shooting out blazing cinders. You find that the concentric circles of the path you have ascended have progressively centered your attention; you find your soul composed, its wayward impulses and desires fade away. The hundreds of images of human activities have brought you back to yourself. But at the summit, there are only the seventy-two figures of the Buddha, where individuality is no longer seeking to assert itself, neither male nor female, pure postures of cosmic compassion, directing your gaze to the teeming jungle, the fiery volcanoes, and the benevolent sky.

Such works belong to a fetishist world, where it is the voices of things and of the earth, air, water, and fire that summon us and lead us.

The Evolution of Animism

In the Christian West the most revered images were canvases that had been animated by a spirit separated from humans and from the world: Jesus printed his image on the veil of Veronica and on the shroud of Turin. Saint Luke was revered as the artist apostle for having endowed the church with a painting of Mary, the mother of Jesus. But he was a fisherman: it was Mary herself who miraculously printed her image on his canvas. Later, in Guadalupe in the New World she printed her image on the mantle of an Aztec peasant named Quauhtlatohua.

When we view the Parthenon in Athens, we are struck by the opposition between the great raw dark rock, whose unformulatable forms were thrust up in some ancient collision of continental plates, and the gleaming white construction erected boldly on top of it, a temple determined by the abstract ideas of geometry and housing Athena, the humanized representation of Athens. In and around the Parthenon, the chthonic forces of Greek antiquity, Thunder, Lightning, Fire, Wind, and Oceanic Tempest, were contained within human shapes. There, Friedrich Nietzsche said, the Athenians contemplated ideal images of themselves, in which all their passions—for power, for law, for domestic harmony, and loyalty, rage, jealousy, vengefulness, and lust—are affirmed and glorified.

The Renaissance rediscovered the Parthenon and its statues. For the next five hundred years sculptors will work erecting in public places all over Europe and then in the European colonies in the Americas, the Indian subcontinent, and Africa statues of conquerors and rulers depicted in action, their limbs held in significant gestures. The statues are narrative, depicting the ideals Western states and religions have conceived in successive battles and in the foundations of cities, nations, and colonies. They put in inert matter the intentions and achievements of vision, will, resoluteness, and courage. Stone and bronze immobilize these heroic postures to demonstrate victory over rebellion, disorder, and time.

During the Renaissance Giorgio Vasari recorded how painters introduced perspective, chiaroscuro, and the painted effects of light, and observed too that artists were seen to enhance or modify the religious and cultural ideals depicted and also to introduce ideas of their own. They came to be seen as creators, for the sublime meanings they put in their works were understood to be projections of their creative sensibility and imagination.

In the fifteenth century technological advances used in navigation greatly expanded the conversion of nature into resources and products. European imperialism launched a political and ideological struggle with other cultures. The denunciation of heretical and heathen cultures and of their deities as empty idols attributed the voices heard in them as voices projected into them by false priests and superstitious peoples.[2] The pidgin word *fetissos* with which the Dutch and Portuguese merchants dubbed objects that West Africans kept out of commerce—and that the merchants found valueless and repugnant anyhow—assimilated such objects with *feitiços*, the appurtenances associated with witchcraft in Europe. In the eighteenth century the term "fetish" was taken up by the Enlightenment theory of primitive religion, to designate inanimate objects to which "primitives" attribute demonic powers; in the nineteenth century it was taken up by Marx to designate commodities whose exchange value greatly exceeds their use value; and in the twentieth century it was taken up by Freud, to designate objects taken by neurotics as phallic substitutes. This succession of anthropological, sociological, and psychological theories declare that these things are really nothing but what physics and chemistry records of them and that human minds "project" intentions and powers into them. The spirit that speaks in things is not the spirit of ancestors, deities, or alien beings, but our own spirit.[3] These theories are animist explanations of fetishism.

The philosopher Georg Wilhelm Friedrich Hegel, in his *Phenomenology of Spirit,* depicts a great historical movement in which the spirit gradually separates itself from the realm of material things, opposes itself to them, and then takes them over and progressively reshapes the material environment with its desires. The mind finds that it can actively predict how things will react and, guided by its understanding of them, can manipulate them and reshape all things according to its will. The material realm comes to no longer appear indifferent to the spirit, opaque and impenetrable; now the mind recognizes intelligible structures and relations everywhere in things. At a final stage, the mind recognizes that the intelligible structures and relations that we recognize in things are structures and relations that the mind itself has projected into them. The things do not resist these projections, but acquiesce to them. At this stage the spirit recognizes not only that it is separate from the material realm but also that the material realm is not something separate from it but is utterly subject to it; the spirit recognizes its absolute sovereignty.

In 1964 Andy Warhol exhibited *Brillo Box* and launched postmodern art. Artists no longer seek to produce sublime objects, separated from all utility, which would exhibit an ideal meaning. They celebrate coke bottles, soup cans, and blue jeans, which had been taken to have simply utilitarian significance by the advocates of the fine arts. Consumers put meaning in them by defining their identity, status, and happiness with the individual collections they amass and display in their apartments, and with which they network with other collectors of blue jeans, dial telephones, and baseball cards.

If animism is pushed to the limit, the actual material substance that is animated, that relays the voice of a spirit, can be arbitrary. Yahweh can speak in a burning bush; the spirit of the contemporary world can speak in a coke bottle or a Brillo box. Words have materiality, they have timbre and pitch, attack, accent, and duration; written words have shape, thickness, size, and color. Postmodern artist Jenny Holzer presents, in commercial lettering or in illuminated neon, not inscribed on things, the political and commercial slogans and catchphrases, cues, watchwords, and passwords by themselves, showing us the messages, the orders, that are incessantly projected about us. The materiality of the words recedes before the agitated and urgent messages being projected upon us by ingenious and calculating human engineering experts and their ironic or provocative variations by Jenny Holzer.

What the artist does, by enshrining these banal and mass-produced things, as also the commercial logos and icons in art galleries at this particular time in the evolution of culture, is, philosopher Arthur Danto explains, show that these consumer products focus and draw our desires for identity, recognition, power, and happiness. The voices of these things are our voices: they are what we have fixed as refreshing, exciting, glamorous, beautiful; they relay the identity and worth we are fashioning for ourselves. Postmodern art represents the moment when artists have become lucidly conscious of themselves and of their activity. It is the moment when—and here Danto invokes Hegel—the Spirit becomes self-conscious of its work of interiorizing the exterior.[4] The moment, we will say, when a universal human animism becomes explicitly conscious of itself.

The specific kind of animism now dominant in our culture—that finds everywhere only the voices we ourselves project into things—would be rooted in a primary urge in humans to mark their presence in things. How frail and ephemeral is our presence in deserts and jungles, in the oceans and under the stormy skies! Would not art first arise in the use of inert material substance—clay, stone, bronze—to maintain enduring the fleeting gestures of life?

The Return of Fetishism

Among the fetishes of Africans and Polynesians, found objects—bones, fossils, crystals, burls—are sheathed in often intricately beaded and embroidered pouches or cases. Objects made may also be recognized as fetishes if they are unique or made on the occasion of some chance auspicious event. In 1907 Picasso came upon African fetishes in the ethnographic collections of the Palais de Trocadéro and was captivated by the uncanny power of their forms and designs. Soon he, Braque, Matisse, Derain, Modigliani, and Breton were acquiring personal collections of them.

If fetishes could be experienced as art, could not artworks be experienced as fetishes? A working artist is not one who has an encyclopedic appreciation of artworks but one who has a passionate devotion to materials and forms that speak singularly to him or her. An artwork emerging in his hands captivates the artist and guides his hand; it goes beyond or goes outside whatever meaning the artist had conceived for it. It beckons him toward unknown paths. Are not artworks so many scattered sites outside the domain of work and reason, in the realm of chance? Are they not so many talismans and omens of good and bad luck? There is a factor of luck that makes someone who diligently studies art talented or not. There is a factor of luck that decides whether any carving or painting works or does not.

There are in the "art world" individuals who have deliberately cultivated appreciation of artworks of many epochs and styles; they make a living as professional critics and expert advisers to curators and collectors. But for most of us spending an hour or two in the Louvre strangely neutralizes the force of the artworks, producing fatigue. It does seem that the church, or the home, that has one painting or a few, with which we live, which unceasingly shape our sense of the important and the trivial and to which we return, is the normal case. Even experts with an encyclopedic familiarity with artworks often end up writing of one painting or one church that has haunted and directed their lives.[5]

The form of animism that Arthur Danto identifies as characteristic of our culture finds only our voices in things. But do not our ancestors, do not the dead, do not other living species continually speak to us, summoning us and leading us? A great deal of our art, from our performances of Greek tragic plays to plays and films of the voices of the victims of recent genocides, make us hear the voices of the dead. A major portion of what we came to call art were rituals and ceremonies celebrating the turning of the seasons and the rebirth of nature after winter or after the dry season, and celebrating our relations with other species—the crane dances of Japan and Africa, the totems marking relationship with patron animals. Today we have become once again aware that we share this planet with innumerable other living things, whose voices summon our attention and must also direct our lives. There are today works that do not simply speak to us of our concepts, imagination, values, and pleasures. There is Robert Smithson's *Spiral Jetty,* Donald Judd's *100 Milled Aluminum Boxes,* Michael Heiser's *City;* works made of light, the *Sun Tunnels* of Nancy Holt, Walter de Maria's *Lightning Field;* and works made of air and wind, musique concrète, the random sound assemblages of John Cage.

Are we not today at a turning point in our consciousness and our lives regarding our relationship with the material universe? Molecular chemistry and astronomy have vastly extended the small environment of tools and constructions where we had refashioned a patch of the universe so that it is embedded with our concepts and values. While a postmodern culture and art celebrates, according to Danto, in all things the human desires, conceptions, and pleasures that we project into them, molecular chemistry reveals the things that are in us—analyzing our agitated and contented bodies into chemical compounds suspended in salt water. Could it be that outside inert matter summons and directs our material bodies in ways that biochemists have not yet been able to trace?

We exist on a chunk of rock and minerals whirling about in empty space, where we see scattered in the dark voids a few other rock planets and stars, concentrations of fiery gases. We have hardly begun to work into our conception of ourselves, our values and our pleasures, the revelation by astronomy that the sun is burning itself out as fast as it can, and that in another billion years all animal and plant life on Earth, now already 4.5 billion years old, will be incinerated by the exploding end of the sun. We shall have to find a new conception of material reality and recognize the destination and destiny to which it summons us.

12

Nature and Art

The philosopher Georg Friedrich Wilhelm Hegel put the question, which is better, a painting of a landscape, or that landscape itself?

> The natural tendency of ordinary thinking . . . is to assume that the product of human art is of *subordinate* rank to the works of Nature. The work of art possesses no feeling of its own; it is not through and through a living thing, but, regarded as an external object, is a dead thing. It is usual to regard that which is alive of higher worth than what is dead. We may admit, of course, that the work of art is not in itself capable of movement and alive. The living, natural thing is, whether looked at within or without, an organization with the life-purpose of such worked out in the minutest detail. The work of art merely attains to the show of animation on its surface. Below this it is ordinary stone, wood, or canvas, or in the case of poetry idea, the medium of such being speech and letters. But this element of external existence is not that which makes a work a creation of fine art. A work of art is only truly such in so far as originating in the human spirit, it continues to belong to the soil from which it sprang, has received, in short, the baptism of the mind and soul of man, and only presents that which is fashioned in consonance with such a sacrament. An interest vital to man, the spiritual values which the single event, one individual character, one action possesses in its devotion and final issue, is seized in the work of art and emphasized with greater purity and clarity than is possible on the ground of ordinary reality where human art is not. And for this reason the work of art is of higher rank than any product of Nature whatever, which has not submitted to this passage through the mind. In virtue of the emotion and insight, for example, in the atmosphere of which a landscape is portrayed by the art of painting, this creation of the human spirit assumes a higher rank than the purely natural landscape. Everything which partakes of spirit is better than anything begotten of mere Nature.[1]

13

Nature

When we go to the sequoia forests, the Andes in Patagonia, the Amazon jungles, the ice continent of Antarctica, when we dive the oceans and visit briefly the coral fish, when we soar on paragliders the winds and the sky, we go to visit realms unmarked by human hands, whose condensations, fault lines, membranes, and veins are not so many categories and relations constructed by us or by our culture. We look upon what is visible to the forty-five hundred species of mammals, ten thousand species of birds, twenty thousand species of fish, the between two and thirty million species of insects that are born and grow by themselves. We look at them and they look at us and we see something of what they see and how.

Nature is seething, creeping, crawling, scuttling, swarming. A hectare of tropical jungle contains one hundred to three hundred species of trees;[1] storms crash upon climax forests in never-logged jungles uprooting vast swaths of root-entangled trees; sunken ships are soon broken apart by underwater storms and encrusted with corals and gorgonians, the continental plates drift and collide. Millions of microorganisms are swarming past us in the transparent air.

To go to nature is to leave stabilized and sedentary existence and enter into movement. Moving with the falling leaves in the autumn winds, ascending the mountains with the mists, drifting down the rivers under the wandering or gathering clouds. Moving with the herds of wildebeests, zebras, and impalas in the Serengeti. Not swimming, only steering with our fins in the ocean surge with the coral fish. To go to nature is to produce nothing, to manipulate nothing, to collect nothing. It is to greet the oryx and platypuses and hummingbirds and moths with passionate kisses of parting. Our minds no longer grasp, collate, appropriate, legislate; they become rushes and rhythms and flows. They join the birds in the sky. "Our own ideas move through the velvet cranial spaces as unpredictably as the passage of herons or the brief flash of a startled deer at twilight," Paul Shephard wrote. "They flit through consciousness, . . . are attended to momentarily, and in a flash are gone."[2]

14

In Touch

Hands fondle the sleek surfaces of a body, feel the warm and supple substance of flesh. The body that the sensual hands uncover and the caresses discover is not the effective, operative body, whose postures and diagrams of movements are oriented toward objectives and manipulate implements. Caresses avoid limbs where the mechanics of bones and striated muscles are salient and settle rather on cheeks and lips, breasts, belly, and thighs. The body caressed abandons its purposive posture, its limbs roll with gravity, its substance offers no resistance. Caresses are not gathering information or aiming at an objective. They do not prolong the momentum of moments and initiatives of the immediate or remote past; they are aimless and repetitive.

Hands tease, ignite eddies of pleasure and torment in other hands, in flanks, bellies, breasts, and lips. The pleasure of the other excites the caressing hands; their pleasure is pleasure in the pleasure of the other. They are moved by the other, by the spasms of torment and pleasure in the other.

Pleasure is presence, fills the present, disconnects the past, obturates the future. Nothing is learned, nothing is gained from voluptuous caresses.

In caresses what we feel is not a sign or trace of something intangible: the psyche, the alter ego, the person. We feel a living, sensitive and sensuous body present in flesh and blood, substantially there, whose reality is beyond any doubt.

Under the caresses the body of another is not expressive, the posture and body kinesics are no longer designating means and objectives in the environment. A face caressed ceases to be a place where questions, appeals, demands, indications, and information materialize in focused eyes, raised brows, lips shaped. The body of another no longer speaks, the lines of its discourse breaking into nonsense and laughter. Vocalizations are murmurs, cries, sighs. Under the caresses the body of another is infantile. Our hands that caress our body are infantile and animal hands.

We caress horses and cats and cockatoos as we caress one another and as horses and cats and cockatoos caress one another and us. We also caress statues, carvings, pottery, trees, wooden furniture, cotton plants, and cotton garments, which do not return our caresses.

Responsibility is experienced when we find ourselves before someone who singles us out, appeals to us and puts demands on us, someone whose needs are important, urgent, and immediate. The sensual involvement with another occurs when there are no such needs, no such demands, when one's hunger and thirst, cold and homelessness have been satisfied. In sensuality someone presents himself or herself disarmed, denuded, needing nothing from me, surrenders himself or herself to me. Not to my deliberating and instrumental will but to my body, to pleasures that are not divided between mine and the other's.

Sensuality is an irresponsible responsiveness to pleasure and discomfort. This responsiveness is immediate, it occurs on contact. It is pleasurable, a pleasure in the caressing hand that responds to the pleasure in the body caressed. In the giving of pleasure the caress is not asking for return, for remuneration. And the aimless and irresponsible caresses are heedless of consequences.

The sensual touch becomes erotic when there is violation of the person of another. There is a breaking down of someone's public and decent presence and functioning in the ordered and regulated roles of the social field. There is a specific excitement in the collapse of posture and wariness, the divestment of functions, roles, and self-respect. In the surrender, the exposedness to another, there is assent to the passivity and the excitement of pain.

The sensual caresses give way to the hold of bodies held tight in support, and to the touch of consolation, which does not effect or promise healing or deliverance but is afflicted with all the grief of the other, accompanying the other who is going no where.

Words are stored in books and computer files and broadcast at large, but first words are addressed to someone who faces and who calls for my attention and requires a response. What gives vocative and imperative force to another's words is the move by which he or she singles me out and faces me.

In facing me another exposes his or her strong and independent but also vulnerable and needy existence. Her susceptibility to the hard edges of things is visible on her skin, his mortality visible in the wrinkles and scars. I do not observe the vulnerability on the face of another and do not construct it by interpreting perceptual data; it affects me immediately. I make contact with her pain and I feel it in my body. My eyes do not look at the sensitivity of her bare skin and its wounds; they flinch, they feel his pain within their gaze. In the handshake of greeting and my arm extended in support, I feel the exposedness and mortality of the other.

The word "tact" designates a light touch, supple and agile, a holding back. It contrasts with the touch involved in the apprehension, appropriation, and manipulation of tangible things and of others. Tact that holds back one's forces and intentions is a sensitive form of receptivity. It's the body in the room that imposes tact! In tactful dealings with another I am aware of his or her anxieties, rage, shame, shyness, and secrecy.

Contact with another in his or her vulnerability and mortality becomes ethical when we make contact with the things the other cares about and worries about, the things that delight—the patch of land on the mountainside the refugee longs to return to, the vibrant warmth of the slum the imprisoned guerrilla loves, the frogs and wildflowers of the marsh the child loves and that is being drained for industrial development.

Contact with the other in his or her vulnerability and mortality can produce the pity that is simply the contagion of misery and debility. And the simple impulse to help, to supply, to cure of itself is a will to power. Tact understands that the other may need and want his suffering in pursuing his or her destiny. The lover needs the anxieties and torments of love. Love is rare because we fear it, knowing that we are never so vulnerable, never so easily and deeply hurt, as when we are in love. The mother has to grieve for her son. The mother has to grieve for the suffering her son plunged himself into in devoting himself to the suffering of his people, committing himself to armed revolution.

In current language we refer to tactful behavior—tactful ways of approaching someone or keeping at a distance from him or her—and especially speak of tactful language. It may seem that the notion of touch in the word "tact" is only metaphorical. But there is a speaking that from a distance makes contact with the heartache, fury, mortification, wariness, and secrecy in someone.

When out of the crowd in the street someone cries "Hey Al!" these words aimed at me reach me, penetrate through my outer appearance, even my socially coded pantomime and role, to me, to everything I can mean by "me." Whenever I do answer, answer for myself, I acknowledge that that is what has happened. Even when I deliberately respond with the voice of my social personage, my office, answer as the dean, the policeman, the father, I sense that the real me has been touched. Contact was made with the reality, the substance, and not only with signs or traces of me.

And my words too touch the other. With everything I say to him or her, I sense how I affect him, trouble her, question, distress, probe, anger, support, amuse, console him or her. I see it on his face, on the quivers and spasms of her skin, on the tightening or recoil of his hands. My words, my responses—not only their content but their tone, their pacing—are moved by the other's vulnerability and distress.

Tact does not simply express respect for the other, as a caress does not simply express erotic pleasure in the other. Respect for the other, ethical behavior, is real, is realized in tact, as eroticism is realized, is real in caresses.

There are words that keep in touch and that touch, and words that caress, warm, sensual words. But both eroticism and ethics especially require silence. In erotic contact, the coherent lines of discourse break up into teasing, nonsense, and laughter. Erotic contact disconnects from the rational language that seeks to establish truth, that is, what is true for everyone; erotic utterances are a private language. Nothing is to be learned from listening in to lover's talk.

In the ethical relationship that makes contact with the other's vulnerability and mortality but also with his or her inmost wellsprings of energies and exhilarations, tact is made of silence. The silence that is listening in, receptivity, sensitivity to what the other feels and dreams. In fact this silence, this listening is in all the words we say; it modulates their tone; it puts forth whatever we say as subject to the other's consideration, judgment, refusal, and assent.

We do succeed in finding the right touch, the right words, the right tone, or the right silences when speaking with someone whose complex situation and confusion we touch. We can see this tact in practice among refugees, among victims of famine and of the plagues, among peasants laboring on harsh lands in uncertain climates. We can see this tact in practice in every friendship.

With words we stay in touch with things. We also recognize and respect those who have long and deep experience with things. Thought that has dwelt long and intimately with a painting by Rembrandt, a temple in Cambodia, a willow tree in one's backyard finds the right words with which to speak of them.

Tact, that finding the right words and the right silences, is not only a relationship with real people; it is also a relationship with things. The language that seeks to make contact and stay in touch with the Colca Canyon, with the baobab plains of the Sahel, with Angel Falls, with the Kalahari Desert, with a hamlet glowing in the Himalayan twilight, finds the right tone and the right silences and is laconic. It is not the website that stores everything anyone has been able to say about them, but poetry or words of a song that keep us in touch with the real things with which we have made contact. Unrestrained garrulousness is as much a lack of tact about things as it is about people. How coming into the real presence of sequoias, opals, fossils silences us.

Part 3

The Sacred

15

Sacrilege

A sanctuary, a sacred precinct, is a dark or radiant place marked out by prohibitions and taboos, separated from the profane sphere. To encounter there the sacred is to encounter the power of what is separated from, what marks the limit of the world of work and reason, of calculation and appropriation.

Sacred precincts are places of sacrifice. Modern world religions, which have striven to rationalize themselves, regard sacrifice—of goods, other animals, even of firstborn children—as traits of primitive and superseded religions. Yet, Henri Hubert and Marcel Mauss argued that sacrifice is the most universal, perhaps the most fundamental of religious acts.[1]

In a sacrifice something supremely precious—our finest harvest and livestock, our firstborn son—is set aside from all use, separated from the profane sphere. What is set apart from all profane use is separated absolutely, definitively, in being destroyed.

In a sacrifice the burning and killing reveals the separate, sacred, power that limits the space of all work and reason. In sacrifice of the food from the harvest the violent, indomitable power of the flames blazes over human works. The knife that tears open the body of the sacrificial victim tears apart the protective hide or skin exposing the writhing turmoil of spilt organs, reveals the violence of a stag or boar taken from the wilderness—the inner violence of its life—reveals anonymous untamed forces in the child. The knife of the shaman, the priest, Abraham reveals the unintelligible core of life and the inner impersonal violence in the composition of living things.

The sacrificial priest leaves the profane sphere to perform the sacrifice and acts in the name of the people. Bringing to him of their harvest and livestock, a beast of the wilderness, or their firstborn child, they participate in his deed. Those who perform sacrifice identify with the victim. The Aztec priests covered themselves with the blood of the sacrificial victims, excoriated them, and pulled the skin of the victims over their naked bodies. And we who consign to the sacred sphere our resources, the game from our hunt, our children, identify with them, identify with the victims. The stag or wild boar sacrificed would have sustained and nourished us. How could we not identify with our firstborn child, sacrificed to the mountain god Yahweh? At the moment of the blood sacrifice, the participants find their identity slashed with the knife. When the fire blazes on the sacred victim, it blazes too on us.

The communication continues in feasts where immense resources are consumed and in saturnalia where we abandon our sense of ourselves and controlling will and find ourselves possessed with the forces of pounding music and dance and with violent, erotic, excessive compulsions, and with the forces unleashed in the forests and rivers by night.

I had gone, one evening, to see an exhibition of photographs taken in rural Poland by a student in the fine arts department. After a while I saw people passing to a room behind the photograph exhibit and joined them. There we saw a man, powerfully muscled and virile, naked, hanging upside down, his feet bound by a rope looped over a hook in the ceiling. Up against opposing walls of the small room there were stands with piles of knives on them—butcher knives, serrated knives, hunting knives. Against the other walls there were stands with guns and boxes of bullets. On the walls there were maps of the country with the numbers of people killed with such weapons in each of the states the prior year. For I think some forty-five minutes we stood around, viewing the man hung from the ceiling and looking at the stands in a kind of brutalized silence. This muscular young man naked and hung upside down exhibited human life at its strongest, now in a position of extreme vulnerability. The provocation, the temptation to cruelty surged in us. We glanced at the piles of knives and guns and the numbers of people cut down everywhere in the country and trembled before the abyss that we felt gaping open about us. Finally one of us took a knife and cut the rope; the man fell to the floor. I heard a student I knew, Andy, who was standing near me, muttering, "The show is not over like that," and saw him spring forward, seize a knife and bring it down with full force into the floor, grazing the arm of the fallen man. But he had thrust so violently that, without realizing it or feeling it, his hand had slipped off the handle and down the blade, which cut deeply into the palm of his hand and fingers. Seeing the blood pouring from his hand, he shook it violently over the fallen man, splattering him everywhere with his blood.

Andy is a musician, and several surgeries later the doctors had not been able to restore the tendons of his fingers enough for him ever to play again. He had a gathering of friends before the first surgery; he moved among us if anything more energized and ebullient than before.

We in the room felt vertiginously our bodies merging with, being absorbed in the man suspended and then fallen before us; his nakedness made us feel the nakedness of our bodies under our clothing, which protects them only from looks, and we felt his vulnerability with and in our flesh. The anxiety that infected us could be broken only by violence, and finally one of us had taken a knife to cut the rope and the man fell to the floor. But Andy felt most the intensity of violence that was not so easily spent and seized a knife to cut the arm of the man at our feet. When his knife slashed his own hand, it realized the identification with the man he struck, and we saw the fallen man everywhere bleeding with our blood.

We felt shock but also exhilaration in Andy's act; he acted in our stead, realizing the temptation to cruelty upon the fallen man we felt and enacting our impulse to release the tension with a cruel release of the blood from our bodies.

In Rio de Janeiro Avenida Atlântica extends along Copacabana beach, with a broad walkway on one side where restaurants and cafés extend their tables and tourists gather, facing the beach and the shimmering silvered ocean beyond. Around the tables about which people are dining, local people circulate selling paper cones of peanuts, small sausages, candies, cigarettes, and tourist souvenirs—straw hats, T-shirts, postcards. One day I was having a sandwich there, and I saw a man approaching and saw that he had both hands amputated, the one at the wrist, the other above the elbow; he was going from table to table badgering the people at their dinners.

As he approached, my eyes were drawn to him, as our eyes are before people with amputated limbs or crippled bodies; the sight of them arouses in us a wondering how they cope, especially if they do not look feeble and despondent but, instead, energetic and enterprising. There is something deeper: it is not in the nature of our bodies to tend always to a state of equilibrium like inert things; the excess energies our bodies generate prepare them for hard tasks and for dangers. But they also launch our bodies gratuitously into obstructed ways and on perilous paths. They drove our ancestors to cross the glaciers from Siberia to North America during the last Ice Age; they drive our contemporaries to descend into the ocean abysses and rocket into outer space. Subliminally our bodies see in amputees a premonition of wounds and mutilations and eventually incapacitation that lurk as possibilities in our most energetic drives.

When the man got closer, I saw that what he was selling, thrusting the stumps of his arms before the diners, were—knives. His performance was in the public space, but his audience had not come there or agreed to watch its demanding and grotesque plot. The diners at the tables were first repelled by another peddler offering them tourist souvenirs, then revolted by the sight of the stumps the man thrust under their eyes. Their revulsion was immobilized by surges of sympathy for not just another impoverished local but a mutilated human being. But then their horror before the sight of his fingerless and handless stumps returned before the sight of the knives he was offering them. Offering them, superficially, for sale, but also as a tourist souvenir of the knives that had cut off his hands, knives for them to use not simply to cut bread or meat but to cut into their bodies.

There are wounds cut open that do not heal. I have not been able to forget this man, his stumps, and his knives; each time I recall him I see him as vividly as the first time and the violent emotions his performance aroused surge again. No doubt some of those vacationers in Copacabana have been able to wipe out the sight of him from their memories. They would have had to stifle the passions, of empathy, of morbid fascination, of cruelty to him and to themselves that he had awakened in them. When he moved on, they would look at him from a distance and remind themselves that they indeed had sympathy for any unfortunate human being. They would recall the times, back home, in their gated communities, when they gave to charities, indeed, volunteered on Saturdays at the information desk of the hospital. They reassured themselves of the kindness of their hearts.

Sacrilege is akin to sacrifice; it is the converse of sacrifice. Sacrilege is a violence that breaks through the taboos and prohibitions that enclose the sacred precinct, to fiercely confront the sacred. Sacrilege encounters the power and violence of the sacred.

It is possible to view sacrilege as a violence directed against the culture and social structure of a people. The Frankish conquerors cut down the sacred trees of the Goths. As the Catholic monarchs conquered the Iberian Peninsula, in the thousand-pillared mosque of Córdoba they contrived a cathedral. Cortés destroyed the Aztec temple in Mexico City and built a cathedral on the site. Unable to dismantle the 4.45 million square meters of stones of the pyramid of Quetzalcoatl at Cholula, the greatest structure ever built on the planet, the Spanish built a church on top of it.

But J. C. Heesterman showed that in India the sacrifices of a society became individual sacrifices.[2] The seven-horse sacrifice performed by high civic officials in the name of the state was no longer performed by the beginning of the Common Era; instead individuals performed sacrifices. In Haiti and in the Brooklyn home of Mama Lola[3] sacrifices are rituals requested and performed by individuals under the guidance of a Vodou *serviteur.*

Sacrilege, the violation of the places and rituals by which individuals perform sacrifices, would then be viewed as a violation of the conceptions with which an individual conceives his or her place in the cosmos and his or her significance and dignity.

Today secularists reduce the experience of sacrilege to the simple will to power of a conquering people. Michel Leiris, curator of the Musée de l'Homme in Paris, detailed in *L'Afrique fantôme* a two-year expedition in the Sahel to enrich the museum, tricking, bribing, and outright stealing ritual and sacred objects from the Kono and the Dogon.[4] Later, Thor Heyerdahl recounts, in *Aku-Aku,* how he acquired by fraud, espionage, and theft the family heirlooms and protective carvings from the hidden family tombs of the last of the Easter Islanders—a book that makes sickening reading today.[5] We see those who practice sacrilege—the curators of the museum in Cairo where the pharaohs are on display, taken from their tombs and stripped of their utensils, weapons, and finery destined for them in their afterlife, their heads stripped of their wrappings—we see the sacrilege as driven simply by rationalist and scientific hubris.

But there is more to it than that. Sacrilege is an intense experience, which involves a boldness to confront the sacred, to do violence with the violence of the sacred. The Frankish priests who cut down the sacred tree of the Goths, Cortés and Pizarro did not think that the deities of the Goths, the Aztecs, and the Inca were nothing; they acted to show the power of the Christian god over the demons of the pagans. The researchers who penetrated into the tombs of the pharaohs or set out to carbon-date the Shroud of Turin pit the power of rationalist civilization over the power of mythical and mystical cultures.

In the high Gothic chapel of the Holy Blood in Bruges in Belgium people kneel at the altar railing and the priest passes back and forth all day giving them to kiss the most sacred reliquary in Christendom—the Holy Grail, a rock crystal vial containing the blood of Jesus, blood of our redemption, brought there from Jerusalem in 1150 C.E. by Count of Flanders Diederik van den Elzas after the Second Crusade. The chapel is built over a much older Romanesque church, whose foundations can now be visited in the crypt. There, looking carefully at the upper backsides of the pillars, one can see hex signs carved deeply into the stone. There, beneath this most sacred shrine, people executed pacts with Satan. What terrifying boldness and loathing seized them as they did so?

Midafternoon. Twenty-eight years ago. It was my first sojourn in Islam. In the mosque of Süleyman the Lawgiver in Istanbul a hundred or so men had gathered to bow and pray; I seated myself in back on the rug and subsided in the cool of the great luminous domed building. Unlike the Catholic churches of Europe and of my youth, the prayer hall is not a sanctuary; Muslims today, as those of the first generation, perform their prayer equally in the desert. At the end of their prayer as the men stood and moved out of the mosque, a youth came up to me and asked where I came from. He was Omar, twelve years old, and he invited me to come see the mosque. A mosque is a place for the works of mercy; in this building complex designed in the sixteenth century by the architect Mimar Sinan, we visited the dispensary of food for the poor, for the distribution of clothing, the clinic for the sick, the school, the dormitory for the homeless. Omar said he would like to practice his English; we agreed to meet every other afternoon, and he would show me the city.

Before leaving to meet him, I would look through my guidebook. One day I suggested we go to the Phanar district where the guidebook had listed a few little-visited sights. This is the far end of the city, just this side of the land walls. It was two in the afternoon in midsummer; people were dozing in the shade of cafés and on divans behind the doors of their homes. We came upon the cathedral of Saint George. It is a brick and stucco church, smaller than any Orthodox church in Pittsburgh or Baltimore. But it is the seat of the Supreme Patriarch of Orthodoxy, cringing here far from the shadow of the Hagia Sophia that for a thousand years had been the largest building in the world. Omar signed that he would wait outside.

Inside, the dim light and walls grimy with the smoke of candles and incense contrasted with the vast luminous spaces of the great domed mosques of the city. Here the central nave would barely hold a hundred worshippers. In front, the iconostasis shut off the sanctuary behind which the Supreme Patriarch would perform the sacraments. It was impossible to appreciate the art of the icons stuck along the walls, for tin sheets covered all but their smoke-blackened faces. I turned to the right wall behind an arcade of brick pillars, where instead of a series of chapels I saw there were big marble catafalques. The covers had raised emblems with ornate crosses on them. These, I thought, must be the tombs of the Patriarchs of Orthodoxy. I walked the length of the series, to the front entrance, and looked back into the central nave, then up the left aisle; there was no one in the church. I thought of Omar waiting outside, and felt something of distaste for this place that he would feel. The last appurtenances of Orthodoxy had ended up here where the city ends.

I moved back to the first mausoleum I had come upon; it was also the most ornate. I looked down the aisle, turned back to the central nave: there really was no one in the church. Then, blood hot in my face, I leaned over and with both hands lifted the heavy lid to peer inside. But inside there was a bronze coffin. I lowered the lid, feeling jumpy and grimy. I looked back into the central nave, at the still closed entrance door. I moved back to the catafalque, lifted the lid again, set it back and lifted the lid of the bronze coffin. The space was filled with a chasuble of once-white silk now yellowed with heavy raised brocadings in black and purple and beaded everywhere with small jewels. Through the thick smell of dust I looked at the head, cushioned in embroidered silk, a dark brown skull showing under what looked liked shreds of dried beef, scabs in the eye sockets, and patches of skin shriveled from the crooked rows of the teeth. I quickly lowered the bronze lid and then the marble cover and slinked out of the building. I greeted Omar, who asked no questions about the cathedral. He suggested we go have some tea.

The enormity of what I had done tormented me for days, for weeks. Most people, if asked to imagine committing a mortal crime, could not really picture themselves doing so, and if they succeed in imagining committing murder, they can only think they would no longer be who they are. The sacrilege seemed to have undermined what I was.

In this forlorn brick and stucco church to which Orthodoxy had been reduced, sacrilegiously violating the mausoleum to find these sacred princes reduced to dusty skeletons, I violently broke with my own Christianity and went outside to rejoin Omar in the sun.

But I had not acted willfully, in an intellectual illumination of the impotence of a two-thousand-year-old politico-ideological institution; instead I had been mesmerized by a force that emanated from the church, the catafalques, like a temptation coming from them. The darkness of the church had invited me in as a place of secrets, the colonnaded ambulatories on the sides beckoned to me, the gilded lids of the catafalques lay there in the afternoon silence of the completely deserted church as there to be opened. As the minutes passed, the temptation emanating from these things intensified such that I could no longer muster the force of will to turn and leave the church. Jean Genet understood this power things have over us, power to captivate us, bewitch us, subjugate us—the jealous husband who one day bought a gun that from then on determined his days until inevitably, inexorably he became a murderer. The Church Fathers have written that there are incitements in things that are so strong that we can resist them only with the help of divine grace, only when the direct power of God intervenes.

What was revealed under the marble and the bronze covers of the catafalque was the unthinkable, unendurable power of death, the death of the Byzantine Empire, of its sacred rulers, of the most powerful intellects that had flourished, of the works and glories of millions of humans, death that defeats and mocks all human reason and all human works, my death too. Revealed that in the imperialism and wealth of Byzantium, a two-thousand-year-old religious effort, intellectual and historical effort, had always been locked in combat with death. A combat that would continue in tomorrow's global mercantile world, where the rain forests, the glaciers, the polar ice caps, the unending stretches of outer space surrender to work and reason, are surveyed, invaded, and exploited by mercantile enterprises. Would continue in me, whose sacrilege has identified me with the sacred victim.

Part 4

Violence

精彩在

16

Material Culture

From time immemorial individuals acquired identity, worth, and community by cultivating character. His or her courage, loyalty, and hospitality gave an Intuit in the Arctic, a Tuareg in the Sahara, a Pashtun in Afghanistan a sense of worth and pride, and made him and her known and respected. From time immemorial also individuals acquired identity, worth, and community by cultivating a craft and skill. To the question, who are you? he and she answered, I am a farmer, a carpenter, a jewelry maker, a weaver of carpets or silk brocade. In ancient times individuals acquired identity, worth, and community through identification with a hero, saint, or sage: to the question, who are you? he and she answered, I belong to the family and clan of the great lord, I am a follower of Jesus or the Buddha. Individuals acquired identity, worth, and a far wider community by taking on and living out religious or moral ideals: they were Hindus, Christians, or Muslims, Enlightenment republicans or socialists.

In our times courage and loyalty are increasingly irrelevant in urban agglomerations where everything is policed and insured. Crafts have been replaced by machines and skills acquired periodically become obsolete. Heroes, saints, and sages have been replaced by celebrities. In our multicultural cities we no longer find our friends and lovers in religious or ethnic communities.

We go on Saturdays to the shopping malls and, from the miles upon miles of counters piled high with mass-produced consumer goods, each of us makes our selection of punk or suburban housewife dress and Brooks Brothers suit or leather jacket; we take home to display in our living rooms where we show our friends our collection of jazz, heavy metal, or classical albums, our synthesizer or stock of vintage wines, our Ducati or antique Duesenberg in the garage. About these things we form communities, networking with other collectors of harpsichord or heavy metal albums, Tiffany lamps or Mauser rifles, pit bulls or Angora cats. The rich acquire identity and worth and membership in elite communities by acquiring residences in exclusive neighborhoods, yachts, designer clothing, and original works by celebrated artists.

TOURIST WORLD
عالم السياحة
EXPLORE
GRANA
CATARACT
JAMAICA
BABADOOL
MISRTOURIST
NUBIANGIFTS
إكسبلور
جرانا
كتراكت العالمية
جاميكا
بابا دول
مصر للسياحة
هدايا نوبية
بسم الله الرحمن الرحيم
قل أعوذ برب الناس
قل أعوذ برب الفلق
قل هو الله أحد

17

Orders

A logo is an object that corporations come to prize obsessively, that controls the movements of consumer masses, without those masses being able to explicate the properties in it that elicit such compulsive movements. A logo is not a sign, designating an intelligible meaning. "McDonald's" is not a synonym for a kind of "restaurant"; it is the Golden Arches, Ronald McDonald, and the Big Mac (which is not a designation of the substances put on the paper plate). Nor is a logo a signifier designating a proper name; nobody knows who McDonald is or was. For an individual the logo is not a fetish as Freud understood it, a substitute for a missing body part; it is the membership badge of the McDonald's global village, the Pepsi generation, the BMW class.

A logo is identified at a glance, without depth or meaning. The letters are an acronym for unknown words. It is the thinnest kind of material object, all surface, its colors primary without nuance. It is a material object that does not derive its value from being singular or rare; it is everywhere, before everyone's eyes, on everyone's lips. An object outside of commerce that regulates the flows of cravings, money, and people in the market, a logo is not a fetishized commodity as Marx understood it.

18

Filth

Feces, urine, vomit, pus, snot, sweat, smegma, menstrual discharges—these viscous substances oozing out of their shapes, expelled from our bodies, spread their decomposition outside. Outside, contact with them provokes retching and vomiting, itself disgusting. Before them the body rigidifies and stands apart, affirming its form, position, and unilateral orientation. Is not the mind's determination to maintain an autonomous individuality driven by the body's recoil from muck and putrefaction?

Body parts removed are repulsive: toenails, scabs, pieces of skin, hairs shed, blood spilt, severed foreskins and clitorises, aborted fetuses, ears and penises hacked off by torturers. A corpse, a whole organism thrown out of the realm of life, is abhorrent; in it the internal violence that extinguished life in it continues in the spilt blood, released excrement and biles, decomposing glands and tissues. Living bodies expelled from the body politic—traitors, outlaws, exiles, refugees, criminals, prostitutes, street kids—are viewed with anxiety and aversion.

Preparing food leaves blood drained, hides and inner organs cut away, husks and peelings of fruits and vegetables discarded. When the restaurants close, the most elaborately blended and decorated cakes, unsold, are thrown into garbage bins along with the leavings and scrapings from pots and plates. Scraps and shavings removed as craft or industrial products are shaped sink into the formlessness of litter. Industry produces toxic air, polluted water and ground. Whatever is used is used up, whatever worn is worn out and turns into refuse. Preachers, politicians, and pedagogues incessantly produce principles, which reverberate into ranting: thoughts thought too much no longer think anything, Maurice Merleau-Ponty noted.[1] Biotechnology, cloning of the most productive livestock and genetic engineering of super-productive plant crops result in a percentage of misshapen and nonviable animals and plants. The high-efficiency industries generate ever-larger numbers of unemployables, drifters who paw at passersby in the corridors of commercial centers.

There are industries that produce nothing but substances toxic to vegetation, toxic or addictive to human and other animal populations. Information technology generates political, commercial, and military disinformation, propaganda, hate speech, and trivializing talk and imagery.

On a living body a missing eye or limb induces tenderness and compassion, but excess body parts—extra fingers and toes, more than two arms or legs, double genitals, genitals of both sexes, an abnormally large head or two heads—arouse horror.

In the shopping malls "consumer goods," resources or implements, are held by or put onto mannequins, then bloated with delirious packaging and soft-core pornographic images. These excrescences are libidinal attractors; marketing works the continual displacement of our lusts upon these posturing racks and hyperkinetic wrappings, which, the commodity purchased or the marketing campaign over, become refuse. Industries frenetically pour out images and flags around a political campaign or catastrophe, a spectator sports event or Hollywood blockbuster, garb and memorabilia to feed a momentary infatuation with a celebrity wedding or funeral. Footages of gibbering consumers in frenetic euphoria are inserted between brief media glimpses of policies and catastrophes in the outside world.

Slogans, buzz words, jingles made from electronic impulses convert reverberation into performative speech acts that are not equivocal or fuzzy; balloons, childrens' toys and sex toys, office and home decorations made from plastic convert resins and crude oil into objects that hold their form, smooth-surfaced objects that do not accumulate grime. But the profligate overproduction of them, their vacuous substance, and the dissolution of meaning in them continually turn them into an opaque and cloying sludge that fills space and time and chokes us with spasms of disgust.

Money—paper intricately engraved, metal disks embossed and polished, made to flow out of the hands of any possessor—defiles and corrupts, is excrement. Filthy lucre.

Slime undergoes strange reversals. Prosperity is the continual wearing out and using up of commodities and the seasonal and daily shredding of styles, packaging, and lures that fuel industry. Buzz words, slogans, chatter, infotainment, the media broadcast of the posturings of uninformed individuals produce public opinion and public consensus. Shed blood, dismembered bodies convert into political and sacred glory. Epidemics, famines, and massacres produced and maintained the feudal order, the order of sovereign nation-states and empires, and today produce and maintain the global corporate order.

In the blood shed in menstruation a fetus is conceived, in an expulsion of blood and slime an infant is born. Flies, spiders, slugs, maggots swarm over the dead body; inside the corpse is teeming with chaotically multiplying microorganisms. In the polluted streets and riverbanks in which we circulate some infantile and libidinous part of ourselves is drawn to the sludge bubbling with toxic and generative energies.

Eroticism crystallizes the grubby flesh with silk and sequins and starched naval officers' uniforms, with coiffures and jewelry, and converts lustful impulses into dances, games, charades. These decompose into seductions, sequestrations, and the voluptuous craving to be violated. In orgasmic pleasure the body loses its posture; its limbs, dismembered, roll about, moved by another, moved by the pleasurable abandon of another; one's fingers probe and penetrate blindly not knowing what they are looking for. Dissolute, possessed, no longer in control of one's thoughts and values, no longer master of one's feelings, invaded by alien feelings, one loses one's skilled and social identity, one's responsibility, one's self-respect. One loses oneself in free-floating and monstrous images, in expulsions of air and energies, in discharges of sweat, semen, vaginal discharges, in the night teeming with stains and microorganisms. Yet this abandon is making love, from which come all the care, respect, compassion, mercy, devotion, grace, admiration, adoration, worship that rebound over beings in the world.

Overleaf: *Figure CFC798770766*
from "The Chapman Family Collection"

19

Fake Fetishes, Disrobed Mannequins

British artists Jake and Dinos Chapman[1] are brothers. For twenty years their artworks have been made working together. Theirs is not a family but a political collaboration, they say: working together undermines the ego.

As we approach the darkened room where *The Chapman Family Collection* is on exhibit,[2] we first notice the smell—fetid smell of wood, grease, and smoke. We enter the room and see thirty-four stands on which carved wood figures are spotlighted. The eyes of these figures have the entranced look of African and Melanesian masks; their mouths often retain the somber and sealed cast of those masks, as if holding back a portentous secret. They, unlike so many contemporary sculptures, do not leap or dance or ascend from their pedestals into the free space; their compressed postures materialize even the surrounding space and condense it in themselves. They incorporate bones, horns, and shells, animal parts returned to mineral inertia. Several are pierced everywhere with coarse nails, materializing curses launched to work their malignant power over enemies.

Then, on what looked like African and Melanesian fetishes, we catch sight of the McDonald's logo somewhere or everywhere. As soon as we notice the legs of a squat figure forming the Golden Arches, the Ronald McDonald mop of red hair circling a serene ivory mask, or the triangular clown twinkling under the eyes, the somber and sealed figure abruptly empties out into shallow euphoric inanity.

Here the corporate logo, no longer luminous in primary colors in the high noon of public space to direct the flows of people and money, is mired in the coarse and opaque materiality of uncured warped wood shaped by unbalanced carvings. It is desublimated in a confused and unintelligible fall. Hamburgers are no longer enticing savories; they are knob heads of grease- and smoke-darkened wood, they are the brains of aborted dwarfs without bellies reduced to legs, they are the cheesy meat-head of the crucified Christian God.

These objects are not visions generated in the secret recesses of someone's singular unconscious; the fetishes and the logos, anonymous in different ways, that here contaminate one another are found in the broad daylight of our world.

The figures show astonishing variety of design and mood. Each is closed in itself, a comet blazing its trajectory in the empty space that we psychologize tritely as the imagination. They show the skill and completion in which we recognize artistry.

That we recognize them as fetishes is the result of chance encounters with African and Melanesian ritual objects—prized for their own reasons by merchants and adventurers who collected them and had neither the language, the experience, nor the initiation required for them to select objects representative of the culture—and chance encounters by artists and by us with ritual objects in European museums. Some of the most powerful forms in *The Chapman Family Collection* were made by chance: the arms of a ferocious figure have pulled apart his abdomen, but the carving was made with hands meeting; this gaping split was made in one night by the drying of the wood.

In the darkened space of *The Chapman Family Collection,* among the anonymity of the logo fetishes, we find we too are deindividualized. We did not make these things, but we join Jake and Dinos in their compulsions. They chiseled in pleasure, pleasure of the outmoded craft of woodcarving, pleasure of the eyes contemplating possibilities of base materiality that condense ferocity and secrecy and emanate vitality and malignancy—pleasures that we now share. What is integrated and meaningful induces a passive and blissful contemplation; what is absurd provokes the active and voluptuous pleasure of laughter. We laugh, with strong, explosive pleasure before the collapse of function, the implosion of meaning, a laughter reverberating back at Freud and Marx, ethnology museums, the first apparition of West African fetishes before the eyes of merchants, laughter reverberating forward at the global market and the logos controlling the outpouring of mass-produced commodities.

The fetishes of *The Chapman Family Collection* do not disport the exaggerated penises and anuses of many African and Melanesian carvings or of previous Chapman works such as *Six Feet Under.* But the orgasmic eye, the eye that is a cunt, from *Mummy and Daddy Chapman,* recurs here, in the cowry shell that is used as cunt and eyes in one figure, and then is multiplied to completely cover the figure. The Chapmans' and our pleasure in these objects, including the intellectual pleasure—excessive, gratuitous, transgressive, convulsive—is libidinal. A dirty pleasure, for there is in these figures exuding an equivocal smell of grease, smoke, and ash, something filthy.

Zygotic acceleration, Biogenetic de-sublimated libidinal model (enlarged x 1000)

An early Chapman work depicted *Blood, Shit, Semen,* and we have to step across the ground strewn with dog turds to see the artists' caravan in *The Rape of Creativity.* The work titled *Zygotic acceleration, Biogenetic de-sublimated libidinal model (enlarged x 1000)* is a more repugnant kind of filth. It is made of twenty-one fiberglass department store female child mannequins, torsos devoid of genitals, with features of bland and unmotivated euphoria. Here they duplicate one another without separating, their bodies joined at their torsos. They are dressed only in new Fila sneakers, though, twenty-one joined into a circle, some of them upside down, they could not even walk ahead. They have adult tumescent genitals in the wrong places, incapable of reproductive sexuality. The libidinal model is of junk products of mass-fashion manufacture, merged into botched products of genetic engineering.

Department store mannequins are made not simply to display the fashion styles seasonally recontrived for multitudes but also to be seen themselves, and seen with craving eyes. Their faces and positions exhibit consumerist euphoria, but they do not communicate avidity for all these vacuous mass-produced commodities without arousing a subliminal lust. In the display windows they are intermittently seen naked, awaiting the promotion of the next fashion line; then their torsos devoid of genitals permit the deviant eyes of passersby to fondle their bodies. *Zygotic Acceleration,* the Chapmans' libidinal model, is of honeyed-skin child mannequins with adult and aroused genitals exhibiting themselves to our eyes. Ethologists have established that to recognize one another many other mammals and birds focus their eyes on the faces like we do. The genitals of these children are in their faces, penis noses and anus mouths, where under halos of glistening hair they are aroused, aroused by being seen by us. Their excitement spirals incestuously around their conjoined bodies.

The indefinitely reproduced vacuous euphoria of their mannequin faces and their disposable plastic insubstantiality has been halted in the air-brushed and satiny perfection of finish Jake and Dinos Chapman have given to these figures. *Death 2003* shows one pink plastic blow-up sex doll fellating another, but to our touch what looks like plastic turns out to have been cast in bronze. In *Sex I* the maggots, slugs, crabs, flies, rats, and carrion birds that are swarming over the shreds of rotting flesh still left on a corpse are cast in bronze. Filth contained, congealed, solidified, sealed, set apart from all sacred or profane utility, valued at thousands of pounds sterling.

Zygotic, Six Feet Under, and *Tragic Anatomies* exhibit products of genetic engineering that gambol about with their dysfunctional bodies and with faces euphoric and radiantly innocent of all their deformities and excrescences. Mass-produced department store mannequins, they are all surface, made to be seen, devoid of the depth of anguish and pain that would release in the viewer the catharsis of terror and pity through which the viewer affirms his or her moral separation from the fascination with naked human monsters. Their luminous polish and their position in a temple of art allow a perverse pleasure to throb in us. We realize that we like them and return to view them again.

We immediately understand the mass-produced and redundant consumerist euphoria and pedophiliac desire that they in their monstrous forms trigger, and trigger now so blatantly in ourselves, and we laugh. Laughter is an interruption and shattering of the concatenation of consequential thought and of progressive practical undertakings and a convulsive and hedonist return of our bodies upon themselves. The peals of laughter hold on to this moment of breakup, experiencing it as a release and as pleasure. They provoke not an ironic laughter, mocking, scornful, or derisive—that heightened self-consciousness that reaffirms the self-possession and presiding rationality of an autonomous individual—but a lusty laughter, as when seeing our laughter in a mirror rebounds into a laughter agitating our limbs and belly and turning into voluptuous abandon.

The laughter that interrupts and decomposes the moral autonomous individuality in us obscurely knows and calls forth a new kind of existence for us. Henceforth the question of what that existence will be hovers before us. The art of the Chapman brothers, radiant and captivating, induces lusty thoughtfulness about the physical objects we are most attached to and the physical substances we are most repelled by.

20

Wallowing in Glory

Violence, we say, is blind, but it is also a vehement craving to know and a glory in knowing. Violence against persons breaks through the garb that masks and disguises and breaks up the discourse that prevaricates and dissembles, exposes pretenses and hypocrisies, making the real drives and emotions show themselves. Violence reveals the strengths and the weaknesses of the adversary, his or her courage and cowardice, his or her pleasures and terrors.

Beyond the unbridled craving in violence to know an adversary or an intruder, there is a craving to know death. Not only the death of others, but our own death fascinates us, and drives the violence that is energized by death, exults in death, glories in death.

Nothingness and Glory

The primary sense of being alive is the exhilarating sense of an organism surging with excess forces. This positive sense of our life precedes and makes possible the negative sense of need, the sense of lacks that periodically open in our organism. As we grow, so many of these lacks open as a result of our compulsion to release our excess energies in gratuitous actions. The positive sense of our life also precedes and makes possible the sporadic negative emotion of fear for our life.

The impulse to act on our own, to be on our own, comes from the positive sense of excess powers within us. We find ourselves confronting beings that resist and limit our powers. There arises the will to accept these limits, but there also arises the will to abolish them. Humans who are powerful and whose existence and forces are perceived as obstacles to our freedom can arouse the hatred that wills they not be there, wills that they not be at all. Every day across the planet innumerable acts of violence are directed against individuals in dominant positions or recalcitrant individuals in subservient positions, against domineering parents, against burdensome children. Innumerable acts of violence are directed against powerful animals of other species, even bears in the wilderness, sharks in the oceans, and eagles in the sky. Inanimate beings, buildings, bridges, boulders can be perceived as obstacles to one's sovereignty and hatred can destine them to annihilation. In 1950 a young Zen monk destroyed by fire the Kinkaku-ji Temple in Kyoto. In his book *The Temple of the Golden Pavilion,*[1] Yukio Mishima set out to understand how this fourteenth-century temple, celebrated as the most perfect building ever created in Japan, could have aroused his nihilist hatred. Violence divines the sovereignty of a freedom finding no resistance or limits, the glory of a solitary star in the emptiness of outer space.

Hatred arouses hatred. My adversaries, seeing in me the wish that they not be there, a will to exterminate them, turn to me eyes in which I see their wish that I not be there, not be at all. In the hatred I see in someone's eyes I see my death identified with annihilation.[2]

The nothingness that others will for the violent one appears to him to be the ultimate limit to his freedom, and to render inconsequential all the other limits that he has overcome. His will to sovereignty can now lead him to set his existence, standing firm on itself by itself, up against this abyss of nothingness. He seeks to make the power of death his own power.

The oldest forms of human society were bands of hunter-gatherers. Hunter-gatherer societies today are individualistic, without masters and subjects. The small groups avoid other groups, avoid territories others have already hunted and foraged over. From time to time their paths cross and they do fight. These battles between nomadic hunter and gathering peoples are not fought to occupy and hold territory or even to seize booty or women. They function, anthropologist Pierre Clastres argues, to maintain the autonomy and distinctiveness of groups.[3]

But in many of these societies, there arose "professional" warriors, associations of men permanently dedicated to war. Clastres researched these warrior associations among the Guaicuru-Maya of the Gran Chaco and studied the research done on the Yanomami, Iroquois, Apache, Huron, Algonquian, Cheyenne, and Sioux. Warrior associations among the Papuans of the New Guinea Highlands and the Ifugao of Luzon in the Philippines have also been well studied. Western observers noted that often it was simple insults that provoked war, and that the wars did not profit the victors, for the defeated only awaited their chance to resume battle and wreak revenge. Wars between neighboring, even related clans and tribes, what we might call civil wars, sometimes continued for generations, sometimes resulting in massive depopulation of both sides.

The act of going off to seek out and confront the nothingness that one's adversaries will for one is an act of a different kind from pragmatic actions that marshal one's forces to move, manipulate, or to destroy beings. Here the warrior may be killed, achieving no material objective for himself. Or even for his comrades, for every violent act forthwith provokes a violent response, unleashing what may well become an interminable chain of deadly ambushes and raids. The act of going off to face death, where one stakes all that one has and is, is an exultation that seizes the warrior and an ecstasy that becomes an end in itself.

The warrior associations are as individualistic as the "primitive" societies in which they take form, without subordination to war chiefs or strategic leaders; indeed they are more individualistic, since each warrior glories in the affirmation of his powers at the brink of nothingness. Among the Guaicuru-Maya, all that the warriors brought back from raids and battles were the scalps of their adversaries. Pride, the pride of men who stand fast and withstand the deadly violence of their adversaries, motivates their battles and ensures that their battles are endless. In his determination and his valor the warrior glories in his physical and moral powers and his individuality set forth before nothingness.[4]

The one who harnesses all his forces to stand in the face of the death to which his enemies destine him may succumb to their hatred and their forces. Then his resoluteness is a momentary stand, his glory in seeing himself standing with his own forces before the abyss an ephemeral glory. His triumph in his own eyes vanishes before their definitive victory over him.

If he survives, then his resoluteness and his glory exist for him. But if now he withdraws, or fights only in the rear, he retreats from his ecstasy and nullifies his triumph. To maintain his courage and his glory, he must push his way to the front again, once again throw all his forces into the confrontation with those who are determined to annihilate him. The ultimate act is that of the warrior who attacks the adversaries' camp alone, equaling himself to the power of all his enemies, where the only thing in his favor is the dumbfounding surprise of his solitary presence there.[5]

The glory of his blazing ferocity in the face of the nothingness that his adversaries will for him exists first for the warrior himself. His audacity and his glory also exist for his fellows in the warrior group. The other warriors honor the courage of the one who confronts, alone, the death his enemies will for him, for they see in it the image of their destiny. Each one, seeing the courage of the warrior in the front line, pushes on to take his place when he falls (or when he falls back to the rear).

"Primitive," that is, individualistic, not internally hierarchized, societies honored the brotherhood of warriors, which does not make itself master of that society and does give them protection—honored warriors with celebrations and festivals recounting their heroic deeds. Then, as states, structured hierarchically, were formed, they enlisted warrior associations to their service,[6] the Hyksos to the Egyptian Empire, the barbarians to the Roman Republic, the Teutonic Knights to the Prussian State, the Janissaries to the Ottoman Empire, the samurai to the shogunate, the Foreign Legion to the French Empire, special forces to contemporary nation-states. States have found tens of thousands of young men willing to fight to the death for the most unjust, deluded, unwinnable causes. Conscripts are motivated by the urgency of defending the society to which they belong or of expanding that society; mercenaries fight for gain, for booty or women; warriors are possessed by the glory of confronting and standing bloody but unbowed before the final limit to individual sovereignty: nothingness. But their craving for warrior glory serves the state, serves those who are not warriors. The willingness of warriors to die in battle is seen to serve the security of sedentary societies and the expansionist ambitions of states.

Corruption and Glory

To realize that we are vulnerable and mortal is not just to tremble with fear or anxiety when we see dangers that threaten us; it is also to suffer. The experience of suffering is not an emptying out, a premonition of oncoming nothingness; it is enduring the reduction of our active powers to passivity, to materiality. In physical pain we are unable to turn away or retreat from the substance of our body; we find ourselves mired in ourselves. We are less and less able to marshal our forces against the resistances of things in the environment. We are less and less able to launch initiatives; an inner passivity weighs on us. We find ourselves less able to plan, to foresee, to call upon our past forces, even to recall those forces. We sense ourselves sinking toward death as toward the limit of passivity, a passivity we will no longer endure, that will undermine our ability to endure.

Suffering ages us, and aging is suffering. The inner sense of aging is feeling the growing weight of all that we have done, all our initiatives taken over the years weighing on us with their inertia. Wounds, scars, debilitating diseases weigh down our mobility. Looking at the mirror we see our body wrinkling, sagging, and stiffening in an advancing rigor mortis.

In a corpse we see an organism that violence has struck, from without, or from within, bacteria, viruses, or tumors having blocked the homeostasis of the body, releasing toxins that will break out of the organs and cells. We see violence continuing, ravaging the corpse from within, breaking out, spreading, befouling, and corrupting all it touches, contaminating the ground with blood and biles, polluting the air with foul gases. Body effluvia, vaginal fluids, semen, mucus discharges, excrement are the everyday evidence of corruption; with spilt blood in wounds but also in menstruation they show my body becoming a corpse. They are filth, the primary locus and meaning of filth. Filth is viscous, spreading, polluting, corrupting, dangerous. After an earthquake, a flood, a battle, nothing is more dangerous to the survivors than unburied or unburnt corpses. In corpses we see death, not as an instantaneous annihilation but as a decomposing, corrupting, polluting violence of nature.

A corpse arouses horror in us, but we are also drawn to it; this guise of death fascinates us. Surges of violence are aroused by the sight of corpses, energized by them, exult in them. A first effect of bloodshed is the compulsion to spill yet more blood. Anthropologists find many societies where every death in the community produces an active initiative to kill someone practicing sorcery in that or in a neighboring community, or to kill someone, just anyone, in that other community. Today, without belief in sorcery, the sight of a corpse continues to provoke the violence that produces another corpse—see our crimes of passion, revenge killings, gang battles, riots, unstoppable wars, torture of captives, capital punishment.

This kind of violence aims not to annihilate but to defile, befoul, corrupt. Innumerable acts of violence aim at seeing the lacerated and broken body of one's victim, where his intentions and will are reduced to spasms in the gore of his body. Here violence is provoked not by the solidity and strength of a being, but by the vulnerability of another. It is fired by the sight of corpses and by the preview of a corpse in a living organism.

What fascinates in the corpse is the violence still at work there, the inexorable power of that violence that spreads beyond the corpse, polluting, corrupting. In the inescapable destiny of every free and sovereign agent about him to become a corpse the violent one sees the sovereign power of nature.[7] He wills to take on that violence of nature, to wield it as his own power. She will make herself not a power sovereign over nature but a power in nature, a power of nature, an infection, a plague. "I would like to find a crime which, even when I had left off doing it, would go on having perpetual effect, in such a way that so long as I lived, at every hour of the day and as I lay sleeping at night, I would be constantly the cause of a particular disorder, and that this disorder might broaden to the point where it brought about a corruption so universal or a disturbance so formal that even after my life was over I would survive in the everlasting continuation of my wickedness," Sade's Clairwil says.[8] Her will triumphs not in the glory of a solitary star in the emptiness of outer space, but in the smoldering glory of sovereign nature.

Our newspapers daily report individuals not only exterminating people they found to be in their way, but seizing vulnerable people and subjecting them to torture. To add torture—holding the victim at the edge of death by blows, electroshocks, or drowning but delaying death to prolong maximum physical and mental pain, making the victim see, making him or her feel the corpse he or she is becoming—compounds every crime, making it monstrous.

When a state maintains torturers in its prisons, it justifies torture by alleging that it tortures uniquely to obtain information from captives. Its intelligence teams know well that when a captive is willing to talk in order to get relief from torture he will say whatever he thinks the torturers want to hear, whatever he can make them believe. The torturers strip their captives naked, drag them over the floor, chain them in twisted positions, cover them with urine and excrement, sodomize them with billy clubs or pile them up naked penis over naked anus, not in order to make them inform but to make them confess. Confess that they are nothing but filth, bodies in decomposition and putrefying, nothing but shit, make them confess that their minds, traumatized with sleeplessness and pounding junk music, are incapable of truth, that what they believed in is nothing but delusion.

The torturer is befouled by the blows and slashes he or she inflicts on the bodies of captives, by the shit and urine spilled, by the blood and biles that splatter the walls and floors of the torture pits. The torturer finds in his body the substance where the body of another can be wounded, where flesh can be mired in pain. She finds the fault lines where limbs can be broken, organs lacerated, sexual organs made into instruments of torture. Everything the torturer says to his victims and to himself is deceit and deception, the information she supplies to the government strategists is invented by the tortured captives, what the torturer professes as her cause and her service is fabrication and delusion.

The torturer sees about himself or herself not the void left by the annihilation of adversaries, but denuded, lacerated, dismembered, burnt bodies. The leer and the laughter of the torturer triumphs over her disgust. The leer and the laughter of the torturer glories in arrogating to himself the contaminating, polluting power of corpses.

The Glory of Moral Integrity

The spectacle of corpses, of butchery and torture, also gives rise to another kind of violence. There are those in the command headquarters of occupying armies, and citizens of states who learn of torturers acting in the name of and with the permission of their state and look upon the lacerated, dismembered, broken bodies of the tortured with disgust and horror. In this disgust and horror, they feel rising from their depths a sense of their core moral decency. Looking about at their fellow citizens they recognize this same horror and disgust and recognize the core moral integrity of their fellow citizens. They take this moral integrity to be the glory of civilized people, a collective glory. The sense of collective moral integrity can motivate turning the state, that institution that claims the monopoly of legitimate violence,[9] into a war machine wreaking violence on its enemies. The wars of our lifetime have exceeded the savagery of all prior wars. But our sense of our core moral integrity makes us believe that we are motivated only by the transcendent splendor of freedom, justice, and peace.

It is disgust of certain acts, more than ethical or religious reasons to avoid or suppress such acts, that justifies for the ruling class or dominant ethnic group the structural repression of other ethnic groups and sexual minorities. This collective sense of core moral decency of individuals and groups enables them to participate in police repression, armed suppression of social unrest, and war. It enables them to participate in what sociologists call structural violence—the economic and political structures that reduce masses of people to destitution, insalubrious living conditions, malnutrition, and disease.

21

The Art of War

From earliest times war has been a major topic of art, depicted in epics such as the Mahabharata, the Iliad, the Jewish Bible, and in architectural monuments such as Angkor Wat, Chichén-Itzá, the Cathedral of Santiago de Compostela, and the Cathedral of Qosqo consecrated as Iglesia del Triunfo. This art depicted the ruler as sublime in himself, absorbing into his destiny the lives of nameless multitudes. It depicted the blood of defeated armies and massacred populations turning into golden radiance about the victorious warlord.

High art worked to justify the slaughter of war when the campaign was not victorious by depicting these deaths as redeemed by the divine order above the anointed lord. It depicted those fallen in battle received into Walhalla, martyrs immediately taken into the arms of God, crusaders likened to the Christian Son of God who laid down his life for the salvation of men. After the French Revolution, it is the nation that assumes the functions of God; the blood of those who die in war pulses through the nation and they live on in its immortality and glory.

Francisco de Goya's set of eighty-three etchings, *Disasters of War,* finished in 1808 but not published until 1863, forty-nine years after the end of the Napoleonic occupation of Spain, is the first great work of contemporary art. They depict close-up men cornered and disarmed and then castrated and dismembered, the infirm and aged unable to fight or flee butchered, children mutilated and slaughtered. The great causes of the war—the Napoleonic armies heralding the Enlightenment advancing into the darkness and superstition of rural Spain, the resistance of the indigenous people and their loyalties, traditions, and values—are invisible; soldiers, peasants, women, and children tear at one another like so many rabid dogs. Goya depicts mutilated corpses covered with flies and picked at by vultures under dark skies, where there is no god above to witness, pity, and redeem so much agony, so many deaths.

The classical art of battlefields and wars depicted, in and through the spectacle of mass slaughter, a transcendent sphere of the good; this art invoked the victorious Alexander, Constantine, or Philip II absorbing the agony and death of the brave and redeeming them with his glory, or invoked a transcendent God pitying, honoring, and redeeming those fallen in battle. With Goya both the glorious Napoleon and the glorious

King of Spain have disappeared; God has disappeared. Their place is taken by the viewer, who in his horror and disgust, feels rising from his depths his core moral instincts, an immanent sphere of the good. Thus, after having been suppressed until thirty-five years after his death, Goya's *Disasters of War,* depicting nothing but unlimited and pointless slaughter, now rose to displace classical art and be proclaimed the great and essential humanist art of our time.

Although Goya's pictures of war have been recognized to be truthful to the point that they have virtually put an end to the classical art that glorified and redeemed agony and death in war, they had no effect on the forces that drive Europeans to war. Europe was soon to launch itself into war again, and twice most of the world with it, war to make the world safe for democracy, war to end all wars.

Louis-Ferdinand Céline depicted the First World War in his *Journey to the End of the Night.* He saw no sign of a pitying and redemptive God above, saw the rank hypocrisy of the values and slogans of the Enlightenment, and the vacuity of the nation put forth as worth fighting for and dying for—nations that were themselves nothing but slaughterhouses, governments that function to soak the land with the blood of butchered and mutilated populations. The war literature written not by those who directed but by those who fought the Second World War and the Korean and Vietnam wars divest war of any shimmer of the glory that the heads of state, along with military chaplains and propaganda filmmakers, project on it; soldiers writing their memoirs documented step by step the progressive descent of an ordinary and decent young man into orgies of degradation.

Does this make us think that humanist art has no power to effect the course of human conduct—or does it make us think that the humanist sentiments it provokes—the conviction of a core moral integrity in us—in fact functions to serve the war industry in our time?

Jake and Dinos Chapman spoke of the secret pleasure of Goya barely concealed in his set of etchings of the horrors of war drawn with such artistic perfection. In the year 2000 they purchased for £50,000 a set of Goya's etchings, and painted grinning clown and puppy-dog faces over the faces Goya had depicted stricken with heart-wrenching pathos. By desecrating Goya's work, the art desperately called humanist in our barbarous age, the Chapmans denounce the public's conviction of their core moral instincts as the principle obstacle to the lucid analysis of the state and of the internal dialectics of military technology.

Part 5

Splendor

The Face of Death

The head is about a third larger than life size. The gray sandstone, if once polished, has returned to the granular surface of stones found in nature. The head is turned down a little; the eyes are closed, a face in composure and emanating serenity. The thick sensual lips are turned in a smile blissful and compassionate. The features are not sharply outlined, if once they were. One sees only the dream of a face in the dense substance of the stone. Slowly I came to realize that it was the face of death.

The thought gradually emerged that I would like to die with my eyes on this image. On this twelfth-century Khmer sculpture. I was surprised by this thought; I had never before imagined the circumstances of my death.

In 1979, in order to show the world that the Vietnamese occupation of Cambodia was a fait accompli, the government announced that Phnom Penh and Angkor were open to group tourism. Although forbidden by my US passport to go to Vietnam or Cambodia, I was able to locate and join an Australian tour. We were ten in the group. We had twelve days in Vietnam and two in Cambodia. The first day in Phnom Penh we visited the Tuol Sleng "Genocide Museum," a former school building used as a detention center by the Khmer Rouge police. We looked at the barren rooms where some twelve thousand people had been tortured and killed. The walls of the corridors were covered with the photographs that were taken of prisoners when they were brought here, often with bruised and battered bodies, and then of them after they were executed. We looked at the rooms piled high with skulls and the commingled bones of now unidentifiable people.

In the afternoon we visited the museum in Phnom Penh, which had been closed by the Khmer Rouge authorities. It was there that I saw the stone head. The handwritten sign identified it as that of Jayavarman VII depicted as the bodhisattva of compassion Avalokiteshvara. While the rest of the group wandered about the museum with the guide and then sat down in the small cafeteria for tea, I remained gazing at this head.

The second day we flew to Siem Reap to visit Angkor. Founded perhaps in 802 C.E., Angkor was the capital of one of the great water kingdoms, where wet-rice agriculture required a vast and everywhere-coordinated system of canals and locks and gave rise to a hierarchized political system, at the top a divinized king. Greater Angkor became the largest preindustrial urban settlement in the world, spreading over 1,150 square miles.[1]

Jayavarman VII (reigned circa 1181–1218) is identified by historians as one of the greatest Khmer kings; he extended the Khmer kingdom to its greatest extent, eastward into Vietnam, northward into Laos and Burma, southward into the Malay peninsula. The temples at Angkor exhibit long friezes depicting the wars of the rulers, the massacres inflicted on the armies they defeated, the tortures inflicted on their captives. The temples built under orders of Jayavarman VII also feature such friezes. But there is this sculpture where he is depicted as the bodhisattva Avalokiteshvara. Many scholars believe that the two hundred gigantic faces turned in every direction on the Bayon state temple he had built depict the bodhisattva of compassion Avalokiteshvara. His people and the people of Cambodia today revere him as the greatest of the Khmer kings.

He ordered and supervised the building of the Ta Prohm, Preah Khan, Bayon, and Neak Pean temples in Angkor. He ordered 102 hospitals and 121 rest houses to be built throughout the kingdom. An inscription found in Angkor reads: HE SUFFERED FROM THE MALADIES OF HIS SUBJECTS MORE THAN FROM HIS OWN. He also ordered that schools to train artists be set up in every city and town. The temples built and sculptures chiseled during his realm evolved distinctive stylistic traits, and art historians see the works made after his death to show a slow decline. What today is known as Cambodian classical dance is the dance elaborated and codified during his time.

Was he the leper king that the Chinese diplomat Zhou Daguan wrote of in his chronicles? A bas-relief in the inner gallery of the Bayon temple depicts the king wrestling with a gigantic serpent, perhaps the representation of leprosy, while his subjects look on. In another scene, some people examine the king's hands; in yet another, he lies ill on a bed. Was this great military conqueror, depicted as a bodhisattva, a leper who lost fingers and toes and whose face turned into festering blisters and necrosis, the choreographer of the most refined dance of Indochina?

A student, Daniel, alerted me to an evening offered by the department of performance studies at the university. Having misunderstood the time, I arrived an hour early. It was All Soul's Day on the Catholic calendar, and the space—a very long corridor connecting two buildings—had been furnished to evoke the Mexican Day of the Dead. There were altars decorated with folk art saints, and even trays offering cookies in the shape of skulls, as one would see in Mexico. People gathered, and the performances began. On the far end there was a pile of white sand—I had almost stepped on it earlier. The space was darkened, and an assistant began dropping drops of water into a glass dish of water; the sounds of the drops of water were amplified. From below the ripples were lit and then projected upon the far wall where they spread in luminous circles. This went on for a very long time, monotonous, and we were absorbed with the sense of time itself, endless time, the time of death. Then we saw a movement in the pile of white sand, and slowly a man rose; I recognized Daniel. He was very slender, naked, and his body sparkled with particles of sand stuck to his body hair. His head was wrapped like the head of a mummy. He slowly rose to full height, and then moved around the sand pile in widening circles until he disappeared behind a screen. He was a dead man, and what we had experienced, not in horror but in awe, an impersonal and strangely blissful awe, was the time of dying.

23

The Emergence of Dance

Hands take leave of the trunk and reach out for things; they clench and grip, push and pull. Tools are solid things detached from the environment and handled. Hands make themselves instruments for taking hold and taking back. Hands poke, probe, punch, strike. Hands drop and toss; arms heave and hurl. They make things projectiles and are themselves ballistic. The feet orient the axes of the torso, they drive the outstretched hands. They understand the predator instincts of the hands.

The hand is the earliest container. A clay bowl is the cupped hand freed from the hand. What the hand has freed from the hand can be passed to another hand. Thus ritual begins.[1]

Hands run over contours and hollows, stroke fur and fungus, rub at gum and stroke gooseflesh, spray water and send waves into tubs and lakes. They uncover, discover, behold. They wonder, ponder, and weigh. They are sense organs. The feet raise, the legs bend. They understand the tenderness of the hands circumnavigating the shapes and stroking the pulp of things.

The feet feel rocks and sand and other feet. The toes, the ankles, the knees, the thighs are organs to feel spaces. Rhythms begin in them and send them on along trails and on open roads. Feet that keep moving find spaces free of paths or destinations. They stand on terra firma, rise into the sky, glide in the water.

The feet understand the dangers—the bruises, the lacerations, the burns that the hands, the torso, the tongue may feel. They understand decline and fall, death and decomposition.

Hands apprehend and are apprehensive. They touch the solid in order to detect the void.

A hand extends to the starving child in a refugee camp, a hand opens tenderly about the fledgling bird fallen from the nest in the storm. The hand stroking the dying one communicates no information and brings no relief and extends no hope, is there only to accompany someone in her or his dying, to suffer and to die with him or her.

***

Exploratory and teleological movements arise out of the periodic and rhythmic movements of all animal life. Fish in the sea, butterflies and antelope in the savannah, birds in the sky dance.

Dance frees the hands from exploratory and efficient movements, the legs from hunting and fleeing. A dance moves in a space emptied of the order of things and of things. The bodies are not afloat in emptiness but move in a space full of light and darkness and warmth and, drawing upon the bottomless repose of the ground, stand, bend, roll, and lie, endlessly everywhere mobile.

The abdomen has no muscles to do anything but hold itself together. It pulses with restless blood, milk, biles, sludges. The bared abdomen does not conceal the heart but becomes heart throughout.

Hands move without results or ends in view. They rise, turn back upon themselves, fold, spiral. Movements draw designs as they move into the emptiness. The empty space dilates and contracts, throbs and engulfs.

One hand gyrates around the other, around the hands of another. Hands tease, ignite eddies of pleasure and torment in other hands, in flanks, bellies, breasts, and lips. The arms turn, drift, float upward. The legs extend the warmth, the pulse, the insistence of the heart.

Dance disengages the bodies from tools, weapons, and paths to move them in the space free of taboos where dance sacred and demonic animals.

In the beginning when humans were hunters and gatherers, they lived dispersed in families and small bands. Like any other species, they had evolved in ecosystems that sustained and nourished them. Paleoanthropologist Marshall Sahlins calculated that in the Stone Age humans spent no more than three or four hours a day in work—foraging, hunting, preparing food, making shelters.[2] In their long leisure hours, they, as in hunter-gatherer societies today, took siestas, played practical jokes, argued, took dips in the river, engaged in every kind of sexual play. They also sang and drummed and danced. Song and dance are not intentional acts but rhythmic vocalizations and movements that go on by themselves, generating refrains and variations, picked up from one another, passed on to one another. Disparities among individuals and hostile initiatives dissolve, initiatives dissolve. People from the adjacent families, and then from other groups that passed, assembled for nights of music and dance. The rhythmic and melodic movements generate a specific pleasure, an impersonal, intense energy, and the ecstatic experience of common humanity. Human society was not first assembled out of fear, but out of collective joy.

Sonia is a thirty-five-year-old black woman from the Dominican Republic whom I met one evening at a gathering of graduate students in Chicago. Beautiful, with a soft feminine voice, she has a full mature woman's body. She, alone of the women in the room, had shaved her head. She made art objects and was auditing some philosophy classes. Her artwork was boxes, miniature theaters, in which she exhibited small found objects. She survived on welfare. We exchanged addresses and exchanged letters occasionally. Two years later I learned from a third party that she had been diagnosed with breast cancer and was terrified at the possibility of dying. Knowing that she had no recourse but to the charity ward of hospitals, I wrote her, but got no answer. Later I learned that she had moved back to her mother's somewhere in Florida.

I saw her again three years later, at a meeting on culture studies in Sydney. She told me later that she had received funding to come. The first day of the meeting, at midday, she did a performance. In the entrance hall of the building, in the center of a circle of academics with Ph.D. degrees and secure jobs, she appeared, naked save for a skullcap of white feathers. To a short piece of Latin music, she danced. It was a shock to see her naked: her left breast was a mound of discolored flesh with no nipple; a thick crooked scar extended across her lower abdomen. It was from her lower abdomen, she told me later, that they had extracted the flesh to reconstruct her breast. We saw the botched work done on a welfare patient who could not pay for a competent surgeon. She exhibited her mutilated body and its ineffaceable scars, the issue of so many months of visceral anxiety over her cancerous and mortal body, and the pain, after surgery and the healing, of seeing her disfigured body.

Now that the healing and scarring of her body were done, she was not turned to the future and the array of things open to her; she was not turned to the intellectual fascination with philosophy and to the making of art objects, as though she had never contracted the cancer. In the present she was in the past, reduced just to her body, the body denuded for breast oblation and then for reconstructive surgery. But today she danced. This body could dance, and in the dance it was transfigured.

Sonia danced at a distance from where we stood, but across that distance her nakedness made contact with our bodies. Her fear for mortal flesh infected us; we felt a visceral sense of the vulnerability of our bodies sheathed with clothing renewed each day. The shadow of death closed in upon us. At the same time, her nakedness revealed her courage and her determination.

Dance: it's movement that is not going any where. This absolute movement, wholly in the present, transfigured her body; in all her nakedness she was wholly beautiful. A beauty that enthralled us and as we glanced at one another, somehow the beauty of our bodies, young and old, thick or thin, muscled or sagging, glowed.

24

Collective Performances

From as far back as we can see into our history and across the planet, rituals, initiations, ceremonies, processions, dances have been celebrated. They were performed, anthropologist Victor Turner explained, to promote and increase fertility of men, crops, and animals, domestic and wild; to cure illness; to avert plague; to obtain success in raiding; to turn boys into men and girls into women; to make chiefs out of commoners; to transform ordinary people into shamans and shamanins; to "cool" those "hot" from the warpath, to ensure the proper succession of seasons and the hunting and agricultural responses of humans to them.[1] These performances were also animal and cosmic epiphanies, and often awesome and terrifying revelations of dark compulsions and cruelties that are unleashed in tabooed places and sacred times. They called down plague and disaster on the leaders of enemy peoples and curses on their children and livestock. Collective performances were also entertainments—people laugh freely at the grotesqueries in Balinese shadow plays, in African rituals, in Papuan initiation ceremonies so lavish in cruelties to the initiates; they laughingly recognize village louts arrayed in fancy costume and spouting pompous declamations; they gossip and feast on sumptuous meals.

But what were the effects of these performances? Scientific anthropologists have dismissed their efficacy to increase the fertility of men, crops, and animals, to avert plague, or to ensure the proper succession of seasons. Claude Lévi-Strauss declared that there is nothing to be learned about the order of the world or the nature of reality from the study of myths.[2] Since the Romantics, anthropologists had especially focused on the function of myths and rituals to consolidate and strengthen a community and its hierarchies. But the myths and rituals also engender heretics, break-off sects, eccentrics, scoffers, charlatans, and profiteers. Anthropologists competent in psychoanalysis sought to show how religion and ritual function to satisfy the individual's cognitive and affective demands for a stable, comprehensible, and coercible world. Lévi-Strauss did admit that shamanistic healing rituals do heal, sometimes, or often, and showed how their basic methods, stripped of mythical cosmology, are in use in our scientific psychoanalysis. But the cultural symbols, models, and paradigms that present to individuals a stable, comprehensible, and coercible society and world also engender internal conflicts in individuals and conflicts with those about them. They clash with the temperament, compulsions, and ambitions of individuals, or they exclude individuals with certain bodies or heredity from a life integrated in itself and integrated into the community. See the rubbish men in Papuan communities in New Guinea, the dacoits or criminal castes in Aryan India, the homosexuals in Christian societies. Clifford Geertz speaks of performers in the Rangda-Barong ritual in Bali becoming permanently deranged by the experience.[3] And individuals resist, neutralize, protect themselves from the visions and forces of collective cultural performances.

Collective performances cannot be understood only from the intentions of the organizers, participants, and bystanders, and from their historical, political, economic, and ideological contexts. A cultural performance closes in on itself, its scenes and movements adjusting to one another, and evolves with its own logic, that of ceremony and festival.

Interpretive anthropology views rituals, ceremonies, and dances, collective performances, and cultural systems generally as symbolic complexes that have meaning, and function to give meaning to human thoughts, feelings, and actions. Max Weber declared that the imposition of meaning on life is the major end and primary condition of human existence.[4] "The essence of public symbolic behaviour is that it is a means of communication; the actor and his audience share a common language, a symbolic language. Broadly speaking this is what we mean by Culture," Edmund Leach explains. "When people belong to the same Culture they share between themselves various mutually understood systems of communication."[5]

Indeed, it is this meaning of cultural symbols that, Geertz affirms, first articulates, generates, and regenerates thoughts. To think is to identify things and relate them with words and other cultural symbols. Further, indignation, a feeling of injustice, of frustration of our expectations and plans, envy, jealousy, triumph—words and cultural symbols make them possible. Words and cultural symbols determine what we laugh over and what we grieve over. Georges Bataille argued that it is taboos that produce extreme emotions, which burst forth through the transgression of taboos. "Not only ideas, but emotions too, are cultural artifacts in man," Geertz declares.[6] This thesis establishes a radical distinction between humans and the other animal species.

With the thesis that cultural performances, and cultural systems generally, have meaning, and produce effects through their meaning, goes the thesis that anthropology understands them by translating their meaning into that system of cultural and linguistic symbols which is the anthropological interpretation.[7] "The meanings that symbols, the material vehicles of thought, embody are often elusive, vague, fluctuating, and convoluted, but they are, in principle, as capable of being discovered through systematic empirical investigation—especially if the people who perceive them will cooperate a little—as the atomic weight of hydrogen or the function of the adrenal glands," Geertz declares.[8]

The anthropological discourse laying out the meanings of a cultural system and then of another, and the comparisons between them and between successive systems, is part of the general modern project that we can call our historical consciousness. The emergence of scientific history and the social sciences in the nineteenth century was borne by the conviction that if we could represent the forms human societies and events took across the centuries, and the relations between them, we would produce an encompassing and integrated knowledge of the meaning of social events and understand their consequences.

The concept of meaning, central to cultural hermeneutics, is, however, not very clear. It is not simply intellectual, conceptual meaning, grasped in conscious acts. Geertz reports that when the Javanese speak of their sense of rituals and ceremonies they speak of *rasa,* a term uniting taste, touch, and emotional feeling with "meaning"—but "ultimate significance"—the deepest meaning at which one arrives by dint of mystical effort.[9]

Yet if, as Geertz asserts, the meanings that symbols, the material vehicles of thought, embody are, in principle, capable of being discovered, they are not fundamentally different from the intelligible meaning formulated in the anthropologists' concepts.[10] Indeed, collective performances have the same function as the anthropological discourse: the Balinese cockfight is for Balinese a sentimental education, Geertz explains, a text that articulates for and reflects to the Balinese the emotions with which individuals are put together and society is built, the look, uses, force of, and fascination with the violence in their highly stratified and ceremonious society.[11] That historical sense that we acquire through the objective representation of the forms of society and culture provided by our social sciences, among them anthropology, prior societies acquired or produced in and through their collective performances.

Geertz did find that most Balinese worshippers, and the priests themselves, have no idea who the gods in the temples are or what the Sanskritic chants mean,[12] and Donald Cordry found that in the vast majority of Mexican masked processions neither performers nor audience understood much at all of the costumes' significance.[13] Is not the concept of meaning here being stretched to designate only the patterns and periodicities of behavior that rituals impose, and perhaps also certain sentiments of group solidarity, awe, or fear? Pragmatists and rationalists, and religious reformers denounce behavior regulated by "meaningless rituals." Barbara Babcock notes that in carnivals and fiestas, fireworks, exuberantly fantastic clothing, patchwork colors, and the multiplication of apparently irrelevant masks and costumes "to the point of indeterminate nonsense" suspend customary meanings. Yet she says that "a surplus of signifiers . . . creates a self-transgressive discourse which mocks and subverts the monological arrogance of 'official' systems of signification."[14] But it is hard to argue or verify that "transgression" is the meaning that shaped these irrelevant and nonsensical masks and costumes. Is not this meaning indeed a cultural construction—of the interpreter?[15]

The concept of meaning does not seem to designate the essential in the spectacularly theatrical Rangda-Barong cultural performance in Bali that Geertz depicts as an endless, inconclusive clash between the malignant and the ludicrous. Its effect is that people of both sexes fall into trance and rush out to stab themselves, wrestle with one another, devour live chicks or excrement, wallow convulsively in the mud, sink into a coma—"an orgy of futile violence and degradation."[16]

Could not somc of the meanings the interpreter detects in cultural performances be absurd, designate nothing but blind pain, or indeed designate a world full of sound and fury, signifying nothing?

Collective performances do not only exhibit effigies and icons; they continue in throbbing, energizing, and transforming music, song, and dance. The interpretive movement of the mind that constructs meanings is different from, and clogged up and deactivated by, the rhythmic and melodic periodicities of movement.[17]

Over and beyond, or beneath, meaning, there is power generated in collective performances. Victor Turner ponders over "what I have often seen in Africa, where thin, ill-nourished old ladies, with only occasional naps, dance, sing, and perform ritual activities for two or three days and nights on end."[18] Collective performances release, in music, song, and dance and also in trance, resources of pleasure, pain, and expression in our bodies and in unconscious processes that are untapped in everyday life.

Equally striking, and equally enigmatic, is the production of splendor in collective performances. Here the concept of meaning breaks down. The Byzantine iconoclasts and the Protestant Reform denounced the splendor of gilt, color, form, music, and sumptuous liturgical processions for obscuring and engulfing the meaning of the religious symbols. While collective performances have been much studied as generating political decisions and spiritual trances and visions, there is little about how they generate splendor, and little about that splendor. A people are transfigured in glorious adornments and movements; their experience as they perform is transfigured; exalted emotions surge in them; their assembling becomes dramatic, epic, cosmic.[19]

Also little understood is how collective performances produce kitsch versions of themselves and then disappear, leaving only sentimental folklore and vapid commercialized entertainment.

Now that anthropologists no longer study tribal peoples to exhibit primitive stages of human cultural evolution, and the agricultural and technical skills they studied have little relevance to our industrial mass production of food and commodities, will not the splendor produced in their collective performances be ever more important to us, in the petulant venality of our mercantile culture?

25

War and Splendor

The first time I went to Rio de Janeiro I arrived with a toothache and sought out a dentist. I asked him if he had ever danced in an *escola de samba* (samba club) in Carnaval. "Oh yes," he said. "You have to, at least once in your life. There is nothing else you can do that compares with it."

The escolas de samba are located in the favelas and, the dentist told me, the poorest people put aside a *real* a week for years sometimes, in order to be able to afford a *fantasia* (costume) and dance in the *Sambódromo.* Carnaval is the people dancing, with anyone, with everyone, in the streets with the neighborhood bands, in costume and with floats of an escola de samba in the Avenida Rio Branco, in the Sambódromo. The young, old, transsexual, infirm, the poorest are transfigured with a glamour more extravagant than that of yesterday's aristocrats or today's celebrities, casting their *alegria* (joy) into the crowds where it spreads and gains momentum. I looked out over the plumes springing over tens of thousands of dancers in the Sambódromo, thinking this has got to represent the total ostrich plume production of Africa, the total peacock plume production of India. The people, a miscegenation of native American, European, and African, bare their physical splendor; even the most elaborate costume is splayed as a shrine about the bared chest, breasts, and dancing buttocks and legs of the reveler. In the great parades the escolas de samba present theatrically with dances and with elaborate floats the Rio Amazonas and Rio Tocantins, the spectacled bears, the golden lion tamarins, and the toucans, the Indians of the Amazon and the outposts of the Inca, the queens of Africa, the *bandeirantes* (slave hunters and prospectors), the *quilombolas* (escaped slaves), the travelers of outer space. Everything—plants, insects, birds, beasts, heroes, knaves—becomes beauty, samba, and alegria. The splendor of individuals, groups, and floats, this glorified humanity that is not graded into rich and poor, powerful and impotent, Carioca or foreigner, dazzles the eyes and stuns the mind, and the alegria that surges is a gratuitous and superabundant outpouring of energies that drowns out quotidian needs and concerns for the morrow.

I arrived the week of the outbreak of the First Gulf War, in which thiry-four advanced countries united in no higher cause than to secure for themselves the sources of cheap petroleum. At the Rio Carnaval, I thought this is the important event on the planet. Its cause, its experience, its effects being as important as, or more important than, military alliances, political campaigns, rises in the interest rates, or downturns in the stock market.

Carnaval in Rio begins in 1845. The Guerra dos Farrapos, the bloodiest civil war of Brazil's history, had ended with victory of the imperial forces. The soldiers celebrated riotously their return, shooting their guns in the streets of the capital, setting bonfires, overturning carriages, and smashing windows. Don Pedro I had been forced to abdicate and ten years later fifteen-year-old Don Pedro II crowned emperor, in violation of the constitution; further, the Republican uprising had sympathies among the people. The following year the government feared the anniversary would be the occasion for another such popular antigovernment demonstration, and decreed that the occasion would be marked instead with masked balls, like those that in France had marked the anniversaries of the Restoration of the Bourbon dynasty. But while the ruling elite danced in their elegant ballrooms, the streets again broke out in riotous revelry, with many common people donning makeshift costumes and dancing in the streets. Candomblé priestesses from Salvador de Bahia happened to be in the city and whirled in trance in their long white dresses; they were subsequently to be a part of the celebration. But over the years this political aspect of the street assemblages and this religious significance of the dancing processions were lost, and Carnaval became all about beauty and alegria. High government officials do not attend, and the now rapidly growing Evangelical churches condemn Carnaval for its sensuality and the licentiousness that is unleashed in the commingling crowds. Carnaval is a locus of conscious ethical conflict—between the ethos of collective joy and the ethics of national development—Rio and São Paulo. As they say in São Paulo, there is nothing in Rio except the beach and samba. As they say in Rio, there is nothing in São Paulo but work and money.

In 1964, in Papua New Guinea, some Australian colonial administrators, remembering the Highland festivals in Scotland their immigrant fathers told of, organized the first Mount Hagen Show. The Papuans, they thought, love body adornment and spectacle; it would be a joy to behold and a joy, for them, to celebrate their beauty: a Carnival in the Pacific answering the Carnaval in Brazil on the opposite side of the planet. They summoned the tribes of the Western Highlands, who they perceived to be living in suspicion and hostility with one another, to come in ceremonial dress and parade together under Mount Hagen. The men came in triumphal war dress and with their weapons. But the Australians organized the Show as a celebration of the end of tribal hostility, a festival of the new Pax Australiana.

Before the Second World War, the Australian colonial administration was very thinly staffed, and no effort had been made to extend control over or even explore the mountainous tropical island. Then the European war extended to the Pacific, and Australians, Americans, and Japanese fought in New Guinea. The Australians enlisted Papuans in their war, and some fifty thousand of them were killed. After the end of the war, as the Australians returned to their colony, they seriously set out to pacify the country. Not to put down armed opposition to them; from their first arrival their guns had quickly showed the Papuans the futility of that. Pacification meant that conflict among the Papuans was not to be settled with weapons, but by recourse to Australian administrators and courts. The million-strong Highlands Papuans, discovered so late, only in 1930, had been a journalistic sensation, where the Papuans were called Stone Age people and savages. When it was discovered that each high valley had its own language—eventually 867 languages were identified—and the societies so individualistic that defense of one's land was up to the individual and his kinsmen and clansmen, it was easy to imagine them as in a constant state of war. The Australian colonial administration did so depict them and made pacification its overriding priority.

In fact Highlands men sought their wives from the neighboring tribe; this exogamy maintained contact and negotiations between the big men of adjacent tribes. When battles did break out, they were so constrained by rules and fought with weapons so ineffective—the arrows that are without fletching are really inaccurate—that it would be rare that anyone was actually killed. If someone were killed, the big men immediately demanded and negotiated compensation. If compensation—in the form of pigs, foodstuffs, and shells—were refused, then the fighting would resume until someone of the opposing side was killed and balance restored. Although sickness that resulted in death was attributed to ancestral ghosts, spirits, or sorcery, most often material compensation was arranged if sorcery was recognized to be the cause.

For the Australians, these last, as for the first white imperialists, Cortés and Pizarro, the colony was seen as a source of gold. Later, of silver, copper, oil, and natural gas. The problem was that the prospectors and miners depended on large trains of native bearers to carry their equipment and supplies, and they found that again and again the bearers would not cross boundary lines into the territory of the next tribe. The boundaries were protected not so much by arrayed enemy warriors as by sorcery. To break down the tribal boundaries, the Australians, as they advanced into areas where gold panning or dredging was promising, decreed that all tribal conflicts be referred to the administrators and the courts they instituted and sent military patrols to punish tribes where conflicts were being settled in the traditional ways.

The Mount Hagen Show was organized to demonstrate, to the United Nations Trusteeship Council and the home government, that pacification had been achieved. And to extend the forced pacification into Highlands societies, since the clans and tribes of the Highlands would be assembled where they could meet and communicate with their respective enemies. It was also organized to bring in tourists, for by now tourism had become a major industry around the world. An airstrip was laid out, several hotels were built, and 850 mostly Australian tourists were flown in.

The Australian organizers had announced prizes for the best costumes, the best drummers, the best marchers, the best dancers. However, as soon as the prizes were awarded, fights broke out between the losers and the winners; the prizes were subsequently suppressed. The Show was to occur every other year, but it proved difficult to get the big men of enough clans and tribes to agree for a date, so that it was difficult for tourists to plan to get there.

Now the government of independent Papua New Guinea, whose army and police are effective only to protect the mines of multinational corporations, supports the Mount Hagen Show to affirm the national identity of the Papuan clans and tribes and to display to the outside world their cultural diversity. This year (2007) there are only some two hundred tourists, most of them flown in for just the two days of the Show by tour companies. After thirty years of independence, Papua New Guinea is judged by foreign chanceries and tour companies alike to be a primitive and violent place; indeed tribal war was raging across most of the Highlands and in the capital, Port Moresby, all the bus companies and even taxis had been immobilized for the past two months by the conflict.

A local man I had come to know points out a member of Parliament. I remark that there do not seem to be many government dignitaries here. My companion tells me that this man is the representative from this district. He introduces me to the Parliamentarian; I congratulate him on the splendor of the Show. He looks down. "It is getting hard to get young people interested in it," he says. I had noted that some participants in the Show were wearing fake kina shells of painted cardboard. The government-sponsored Show was beginning to produce its kitsch offspring. Later I think the Parliamentarian was thinking of the situation in Port Moresby and the coastal towns. I remembered that in Lae the arriving tourist is rushed by airport police into a van with steel plates bolted over its body and heavy steel mesh over its windows; three armed soldiers are seated in the van as it speeds through stop signs to the tourist hotel. A government report had found that fully 50 percent of the inhabitants of Port Moresby live by theft.

Before the Second World War, the Australian colonial authority had refused to grant permits for missionaries in most of the territory; its military patrols were too few to protect anything but the places where mining companies were prospecting and dredging. After the war, when Australia set out to regain control of its vast colony, the missionaries were seen as effective and permanent agents of pacification. Catholic, Lutheran, Baptist, Seventh Day Adventist, and other fundamentalist missionaries quickly spread across the whole island, building churches and schools and clinics. They are the principal source of cash wages for most villages. Everywhere the missionaries enjoined their parishioners to demolish the men's house, and husbands and wives to live together as nuclear families. They ridiculed the taboos that had prohibited sex for five years until a child was weaned, and that limited most women to two children in their lives. The missionaries staffed clinics and inculcated hygiene; the population quickly doubled. Occupancy of the narrow fertile valleys in the mountainous Highlands had always been the principal motive for conflict; now all the valleys are overpopulated. The export prices for what agricultural products and coffee are raised continue to fall. Young men go down to Port Moresby, the coastal towns of Madang and Lae, where the great majority do not find work. The foreign gold, copper, silver mining companies, the US InterOil refinery, and the Australian Gas Light Company laying a pipeline to bring natural gas from the Highlands across the Torres Strait to Australia are state-of-the-art high-tech projects that employ very few local people. With the abolition of the institution of the men's house, these young men have not been acculturated with the loyalties, rituals, and duties and obligations of their tribe, and the village and clan elders have lost their authority over them. In the Highlands communities men who lie in wait for travelers and kill them were feared rather than admired; they did not become respected leaders, organizers of exchanges, and orators, did not become big men, but "bad men." Now the term is "raskals." Those who return to the Highlands do so to plant marijuana, coca, and poppies for the smugglers in the coastal towns. Some of these earn enough money to buy guns. The tribal wars that break out now result in far more deaths and the tribal big men are less able to negotiate compensations.

At the Mount Hagen Show the marching groups bearing the now obsolete and purely ceremonial weapons, the presence of representatives of the government, and the large contingent of police and army affirm that henceforth violence is the monopoly of the state. However, the Westminster-style parliamentary government set in place by the Australians does not succeed in establishing political parties with national or provincial programs for development or even, malaria researcher Dr. Ivo Müller explained to me, for public health. The 109 Parliamentarians are in effect tribal big men, working to divert some of the national budget to enrich themselves and their tribesmen. Corruption is rampant in the ill-trained and ill-paid army and police, who readily sell their guns to fellow tribesmen in times of conflict and allege that their outposts had been raided. During the last elections (2002), there were known cases of politicians arming their supporters; six of the nine Highlands electorates were invalidated by the High Court because of violence and intimidation. Once again individuals and clans take responsibility for exacting compensation for or avenging aggressions done to them.

With the dwindling number of tourists in attendance, the entrance fee for tourists has been increased from $30 to $100. Coca-Cola advertised itself as a sponsor this year, meaning, I suppose, that it contributed some money. The participating tribes are given 5 kina—about $5—per performer to help in transportation costs. But some groups have come from the coastal towns of Madang and Lae, and even some from the outlying Bismark Archipelago. So they do not come for the money, but for the experience. I greet a doctor I had met in Madang; he had long practiced here and retired to New Zealand upon Independence. He tells me he had attended the first Mount Hagen Shows in the '60s. I ask him how this one compares with the first ones. "Oh, it's much bigger now," he says.

Despite what the Parliamentarian said, there are indeed young men marching in the Show; the majority of the men are young. You watch them, holding on to archaic, but fearful weapons, chanting war chants the length of the day, and you think that this Show, far from demonstrating the pacification of the Highlands, celebrates a warrior culture, which continues in new dimensions in independent Papua New Guinea. In these marching groups of men without leaders zigzagging across other groups in the field, you see what warfare was to the Highlands peoples, where battles were fought without leaders or strategies, each warrior darting and shooting his arrows where he could, exposed to volleys of arrows and spears, exposed not only to cunning and hostile humans but also to supernatural powers and the weapons of sorcery. Battles where no territory was taken, nor women captured or wealth plundered. You are, despite all your ethics and your civilization, enthralled by the vision. You feel in your throbbing legs and arms the vulnerability and audacity of their bare legs and ballistic arms, not protected by camouflage and body armor but adorned with the most ostentatious plumage. The exultation of their battle cries and chants echoing across the mountains invades you and escalates in you. To the rages and triumphs of today are joined hopes and despairs, terrors and audacities, rages and triumphs from across thousands of years past. They are transfigured with a splendor that closes in upon itself and expands with its own logic.

These extravagant headdresses of plumes, these shell necklaces, these boars tusks and wigs were not only arrayed for war; they were donned for all the decisive events in the lives of individuals and community—for births, for initiations, for deaths, for the great pig feasts to which surrounding tribes, even enemy tribes, were invited. Traditionally, Highlands young men did not have obligations to share in the work of building houses or work in the fields until they are married and set up their own household. Before marriage, young men have few interests in gardening, pig raising, payments or exchange; they spent their time in group festivities and displays.[1] All-night singing and courting parties were frequent; young women dress in their finery and invite young men who come unrecognizable in the extravagance of their facial painting and body decorations, singing long-rehearsed songs in the falsetto voices of birds. The Huli young men, wearing their red wigs decorated with flowers, their faces painted red and yellow, wearing the iridescent blue breast shield of the Superb Bird-of-paradise over their bodies painted red, spent a year parading across the whole territory. They long to be like birds, fleet, brilliant, untamed. Today too one cannot wander in the Highlands without coming upon groups gathering in splendor for such events.

In the Mount Hagen Show the Highlands tribes assemble with all their different languages and cultures and conflicts and also with their past ordeals, combats, and triumphs and face the present and the uncertain and menacing future before them.

> Anyone who manages to experience the history of humanity as a whole as *his own history* will experience in an enormously generalized way all the grief of an invalid who thinks of health, of an old man who thinks

> of the dreams of his youth, of a lover deprived of his beloved, of the martyr whose ideal is perishing, of the hero on the evening after a battle that has decided nothing but brought him wounds and the loss of his friend. But if one endured, if one *could* endure this immense sum of grief of all kinds while yet being the hero who, as the second day of battle breaks, welcomes the dawn and his fortune, being a person whose horizon encompasses thousands of years past and future, being the heir of all the nobility of all past spirit—an heir with a sense of obligation, the most aristocratic of old nobles and at the same time the first of a new nobility—the like of which no age has yet seen or dreamed of; if one could burden one's soul with all of this—the oldest, the newest, losses, hopes, conquests, and the victories of humanity; if one could finally contain all this in one soul and crowd it into a single feeling—this would surely have to result in a happiness that humanity has not known so far: the happiness of a god full of power and love, full of tears and laughter, a happiness that, like the sun in the evening, continually bestows its inexhaustible riches, pouring them into the sea, feeling richest, as the sun does, when the poorest fisherman is rowing with golden oars! This godlike feeling would then be called—being human.[2]

This historical sense that Friedrich Nietzsche here invokes is not produced by an intellectual operation of constructing linguistic representations of the forms and meanings of past social structures, inventions, and conflicts; instead it is produced by a return of the passions that created them—the grief over suffering and mutilation and killing, over lost causes and defeats, over ideals one has betrayed, the courage that has endured all these, the hope that opens long-range horizons, the sense of honor that despises self-interest and cynicism. It is not the individual that willfully constructs these feelings in himself; the ancient passions themselves return.[3]

Nietzsche's oldest and most fundamental source of his doctrine of the Eternal Return was his conviction that our emotions are not simply excited by the stimuli and cultural symbols at hand in our environment; they are transhistorical and animal in us and the most archaic emotions can return in the socialized modern man. It was the conviction that governed his first book: the conviction that he, and not the academic literary critics, understood the Greek tragedies because the passions of Aeschylus and Sophocles pounded in his heart. In the late nineteenth century of scientific, industrial, and mercantile Europe, he saw men and women in whom the instincts and passions of hunters and gatherers, of warriors, of sixth-century-B.C.E. sages, and of Dark Ages saints return.

Nietzsche thinks that these ancient passions return in their full force when the representations a people now make of themselves no longer elicit them. Sigmund Freud found that when anxieties and cravings that date from some trauma or from infancy are brought to the light of consciousness, fixed in conscious representations, they lose their force to drive the individual. When those conscious representations fade out, or are lost in another traumatic event, then the old anxieties and cravings return. In Nietzsche's conception, when a people represents itself as wholly civilized, when its economy, political system, and ethics present only occasions for civilized behavior, here and there the old instincts of hunters and of warriors return; when it represents its future as that of consumers in the global mercantilist economy that, here and there, the instincts of sages and of saints return. More profoundly, the instincts and emotions of our animality return. Behind instincts and emotions that are generated by culture Nietzsche found the wolf, cave bear, camel in us; Zarathustra's overman has the instincts of lion, serpent, and dove crowded in his soul.

Nietzsche sees in the specific historical sense acquired in collective performances the production of creativity. In the crowded favelas of Rio full of immigrants from the Mato Grosso and from the Amazon, former slaves from Africa and wanderers from the old Inca provinces, invalids, old women, desperate lovers, martyrs of vanquished causes, defeated guerrillas, but also men and women who descend to the heart of the *cidade maravilhosa* (marvelous city) and make it theirs, dressed in the garb of the old emperors and aristocrats but still more glamorous, pouring their alegria over the city and upon strangers from far-off lands—what is produced is splendor. The return of so many nonintegrated and conflicting passions, crowded together such that each intensifies the others, produces the discharge of excess forces, which are not channeled into economic or political projects but discharged without recompense, released gratuitously. It is this collective situation that is creative of splendor, for splendor is excessive and gratuitous.

The men and women marching, dancing, filling the air with cries and chants under Mount Hagen, are splendid with the nacreous shells of mollusks, the skeletons and fangs of serpents, the tusks of boars, the teeth of flying foxes, the plumes of birds. It is not humans who invented splendor. Since our brother apes do not sing, and indeed virtually no mammals sing, some anthropologists have speculated that we must have picked up song from birds. Contemplating the spectacular plumage and six-month-long dances of the Malay Great Argus pheasant, Charles Darwin conceded that he could not find any functional meaning in these excesses of splendor.[4] Everywhere humans have observed the dances of antelope, sea lions, emperor penguins, ostriches, pheasants, butterflies, crabs, understood them in their own bodies, and taken them up—dancing crane dances, impala dances, oryx dances. " 'O Zarathustra,' the animals said, 'to those who think as we do, all things themselves are dancing: they come and offer their hands and laugh and flee—and come back.' "[5]

We shall not define with one concept the splendor that glitters and resounds under Mount Hagen, in the liturgical processions in Byzantium and the high mass of Mediaeval cathedrals, in the Negara, the theater-state of old Bali, in Carnival in Rio de Janeiro—in the plumage and dance of the Great Argus pheasant, in the sun's gold spread over the blue oceans, in the fisherman rowing with golden oars. But is not the drive creative of splendor Nature in our nature? We are mesmerized by beauty as birds-of-paradise are mesmerized by their glittering plumes in their courtship dances; we create beauty as in the primordial ocean mollusks create the iridescent colors and intricate designs of their shells.

You are seated on one side of a baseball field, behind a chain-link fence. There are dozens of armed soldiers among you. On the three sides of the field there are banks of earth, on which thousands of local people are seated. The drums pound, and you see a phalanx of men entering from the far corner, marching with high steps and chanting warrior songs and cries. They are marching ten abreast, linked together with adjacent men each holding on to the ten-foot-long lance, spear, bow and arrows, or battle axe between them. They are wearing multicolored woven aprons in front; behind, bunches of long cordyline leaves, green, yellow, and cherry red, like the tails of great birds. On their bare chests they have hung gold-lipped pearl kina shells cut in the shape of half-moons. Their torsos and limbs gleam with oil; their eyes are flashing in their faces that they have painted in primary colors, red, yellow, blue, black; boar tusks have been inserted in their nostrils, flying-fox teeth hooked around the rims of their ears; some wear shell disks hung from their nostrils, covering their lips. They wear woven caps tight on their heads, in which they have affixed rows of flowers and small shells and bands of gleaming green beetles. On top, over a band of filmy cassowary and bright-colored parrot feathers, they have spread crowns of golden plumes of the Cendrawasih, the Raggiana bird-of-paradise, the multibanded two-foot-long plumes that decorate the heads of the King of Saxony bird-of-paradise, the three-foot-long black or white plumes of the Sicklebill, Stephanie's, or Ribbon-tailed Astrapia birds-of-paradise. Then the women enter the field. These are not the young women exhibiting their beauty and sensuality who parade in festivals, carnivals, and fairs in Europe, North America, the Caribbean, and Latin America. They are the matriarchs of families and clans, their bare breasts covered with dozens of elaborate shell necklaces of all sizes, colors, and shapes, necklaces of the foot-long bailer shells, twenty, thirty pounds of them—their bridal wealth. They advance in swaying dance steps, their yellow crinkled grass skirts swirling to the tinkling sounds of the bells they have fastened about their ankles. They too wear headdresses of cuscus fur, eagle feathers, and bird-of-paradise plumes often much higher and far outspread than those of the men. The tourist cannot help thinking of the immense slaughter of birds-of-paradise, but in fact the headdresses are heirlooms carefully wrapped up and stored from year to year, even from generation to generation, as are the tons of seashell necklaces that are the bridal wealth of the matriarchs. The women pound the narrow drums they carry and chant songs extolling the proud history of their clans, their migrations and victories in battle. Another phalanx of warriors enters the field, marching with high steps. Then another. More flanks of women dancers.

The Huli wear huge wigs grown from their own hair, overhanging domes decorated with red and yellow flowers; the men march with high steps and then face one another very close up and dance in leaping ever-faster steps. The Asaro have covered their bodies with white clay and wear large globes with grotesque faces over their heads. A group of young men wearing only grass skirts and bearing no weapons from Admiralty Islands dance in the center of a circle of drummers. Here there are youths painted black with the white lines of skeletons. Some groups are performing charades: someone dressed in whiteskin's clothes cracking a whip over men laying railroad ties. There are small groups doing comedy acts.

The men march, the women dance all day in the thin air of the mountain sky; at the end of the day they withdraw to makeshift huts to take off their adornments, eat, and sleep. Behind, over the gloomy mists of Mount Hagen, a rainbow suspends its luminous colors.

The following day you arrive at sunrise to watch them don their adornments again; when they are finished they bunch into groups chanting and dancing with leaping steps while waiting their turn to enter the field. Your eyes swim in the swirling grass skirts and the glints of sunlight flashing off the shimmering multicolored plumes, your heart and muscles pound with the stamping feet and drumbeats. Their chants echo across the landscape, throbbing on the flanks of Mount Hagen, that second highest volcano in the country, pounding too in your chest; you feel your blood hot and surging with the exultation of two thousand men and women, of 125 tribes, zigzagging back and forth like slow-motion bolts of lightning across the crowded field under the magnesium-white sun.

Notes

Chapter 4

John Berger, *The Accidental Masterpiece* (New York: Penguin, 2005), 119.

Chapter 5

1. Hans Prinzhorn, *Artistry of the Mentally Ill: A Contribution to the Psychology and Psychopathology of Configuration,* trans. Eric von Brockdorff (Wien, N.Y.: Springer-Verlag, 1995). Second German edition, *Bildnerei der Geisteskranken,* ed. James L. Foy (Wien, N.Y.: Springer-Verlag, 1995).

2. Four years after publishing his book on asylum art, Prinzhorn published a volume on the art of prisoners: *Bildnerei der Gefangenen* (Berlin: Axel Juncker, 1926).

3. Jean Dubuffet, "Art brut in Preference to the Cultural Arts," *Art & Text* 27 (1988): 33.

4. Jean Dubuffet, Conversation with Jean Dubuffet, August 1976, p. 24. Cited in John M. MacGregor, *The Discovery of the Art of the Insane* (Princeton, N.J.: Princeton University Press, 1989), 303.

5. Paul Éluard, "Le genie sans miroir," in *Oeuvres complètes, t. II* (Paris: Gallimard, Bibliothèque de la Pléiade, 1968), 786.

6. Ananda K. Coomeraswamy, *The Transformation of Nature in Art* (New York: Dover, 1956).

7. Salvador Dali, *The Secret Life of Salvador Dali* (London: Vision Press, 1968), 349.

8. John M. MacGregor, "Marginal Outsiders: On the Edge of the Edge," in Simon Carr, Betsey Wells Farber, Allen S. Weiss, eds., *Portraits from the Outside: Figurative Expression in Outsider Art* (New York: Groegfeax, 1990), 12–13.

9. Roger Cardinal, "Figures and Faces in Outsider Art," in Carr et al., *Portraits from the Outside,* 26.

10. Quoted in Colin Rhodes, *Outsider Art: Spontaneous Alternatives* (London: Thames & Hudson, 2000), 60.

11. John M. MacGregor, "Marginal Outsiders," in Carr et al., *Portraits from the Outside,* 12.

12. Quoted in Marcus Field, "McTribal: How the Chapman Brothers Carve up Culture," *ArtReview* 52 (February 2003): 43.

Chapter 6

1. Martin Heidegger, *Being and Time,* trans. John Macquarrie and Edward Robinson (New York: Harper & Row, 1962), 79–80; and "Building Thinking Dwelling," in *Poetry, Language, Thought,* trans. Albert Hofstadter (New York: Harper & Row, 1971), 145–61.

2. What we call nature are nature reserves, areas we institutionally set up as recreational areas, with staffs to fight forest fires, feed the deer during periods of heavy snowstorms and cull them when they become too numerous, and reintroduce species that have disappeared from the area. What the natural sciences and life sciences call "nature," Heidegger explains, is a secondary, derivative, and abstract representation of our environment, our world, our habitat. (Heidegger, *Being and Time,* 147, 412–15)

3. Geertz emphatically rejects the kind of positivist theory espoused by Bronislaw Malinowski, according to which religion is a collection of magical pseudoremedies and assurances that illness will be cured and the dead reborn, if not in this world in the supernatural world. "Over its career religion has probably disturbed men as much as it has cheered them," Geertz points out, "forced them into a head-on, unblinking confrontation of the fact that they are born to trouble . . . With the possible exception of Christian Science, there are few if any religious traditions . . . in which the proposition that life hurts is not strenuously affirmed." (Clifford Geertz, *The Interpretation of Cultures* [New York: Basic Books, 1973], 103)

4. Ibid., 106.

5. Ibid., 119, referring to Alfred Schutz, *Collected Papers,* vol. 1, *The Problem of Social Reality* (The Hague: Martinus Nijhoff, 1962), 226ff.

6. Ibid., 118.

7. Ibid., 110.

8. Ibid., 94–96.

9. Concerning Aloïse (Aloïse Corbaz), who was interned in the Clinique de la Rosière at Gimel in Switzerland with a diagnosis of incurable dementia praecox, Jean Dubuffet wrote, "She had been cured for a long time. She cured herself by the process which consists in ceasing to fight against the illness and undertaking on the contrary to cultivate it, to make use of it, to wonder at it, to turn it into an exciting reason for living. The wonderful theater she was always putting on—that incessant talk, incoherent and hardly intelligible (she made it unintelligible on purpose)—was for her an unattackable place of refuge, a stage no one else could get on to, where no one could reach her. It could not have been more ingenious, nor more convenient. And with her great talent, her great inventive intelligence, she smoothed and perfected this theater, to astounding effect. She loved to astound. She had worked out little by little a technique which enabled her to do it with great ease, sparing herself trouble (thanks to her unintelligible elocution which dispensed her from always having to hit on new finds.) She had discovered the realm of the incoherent, she had come to realize the profusion of fruits that it can yield, the ways it opens up, the lights it turns on; she was enam-

ored of it and thrilled by it, never ceasing to wonder at it. But mad, certainly not. Quite lucid, I am convinced, withdrawn into the so very ingenious shell which she had devised for herself." (Letter to Dr. Jacqueline Porret-Forel, cited in Michel Thévoz, *Art Brut* [Geneva: Skira, New York: Rizzoli, 1976], 133–34)

Chapter 8

1. Antony Gormley was born in London in 1950; he studied archaeology, anthropology, and art history at Cambridge University. Then he went to India for three years, where he was initiated into vipassana meditation. He returned to England and studied art at Goldsmiths and the Slade.

2. E. H. Gombrich in John Hutchinson et al., *Antony Gormley* (London: Phaidon Press, rev. ed. 2000), 118. Gormley said that Rodin's first publicly exhibited work, *The Age of Bronze,* exhibited the inner space of the body. "He had a Belgian soldier who posed for a year and a half and nearly went mad. He is looking up and is sort of holding a spear but isn't. All the attributes have been taken away that identified it with the 19th allegorical school and it was replaced by an idea about inner space" (Gormley interview by F. David Peat, 1996, www.fdavidpeat.com/interviews/gormley.htm). This work when first shown at the Paris Salon in 1877 astonished people with the extraordinary exactness of the muscle structure of the model, and critics accused Rodin of having used a cast of a live model. To refute these charges, Rodin deliberately made his next sculpture larger than life size. Gormley, however, is struck at how the sculpture shows the model, exhausted by the year of posing, turned inward in an awareness of the inner space of his body. Could this work have provoked Gormley to make an actual cast of his body?

3. The idea of being mummified or buried alive, zombification, is horrifying. In Poggibonsi, Italy, a young man who had volunteered to be cast for Gormley's work was fainting from anxiety and had to be cut out of the cast after a few minutes.

4. "One of the earliest lead pieces I made was called NATURAL SELECTION, twelve natural objects, twelve 'man-made' objects that evolve in a morphological progression in a provisional dialogue between the egg and the phallus. And I really liked it when one of these exploded . . . because it had a fruit inside, and the fruit inside decomposed—I kept the telegram I got from the art gallery because it was so marvelous, saying: 'CONTACT US IMMEDIATELY FRUITS AND VEGETABLES EXPLODING.'" (Hans-Werner Schmidt, *Gormley Theweleit* [Bielefeld: Kerber Verlag, 1999], 63)

5. Gormley, interview by Peat.

6. Antony Gormley, "Learning to Think" (lecture, online at www.antonygormley.com, accessed 2007).

7. "Sculpture . . . is a small and inert catalyst, a bit of matter used to catalyze your sense of being immersed in light and matter. Through the work you may become aware of the breath passing through the channels of your nose, or

the weight running through your knees. Aware of the world that you inhabit and your aliveness within." (Gormley, interview by Peat)

8. Ibid.

9. *Testing a World View,* 1993.

10. *Edge,* 1985.

11. *Pore,* 1988.

12. *Learning to Think,* 1991.

13. *Allotment,* 1996.

14. It was later moved to De Panne on occasion of the 2003 Beaufort Belgium, then to 3 kilometers either side of the tide line on Crosby Beach on Merseyside in England in 2005.

15. Gormley, "Learning to Think."

16. "I think of sculpture as something coming up from under the earth, becoming as we all are earth above ground, but retaining a feeling of having been hidden and then revealed, a revealed energy still embedded in matter, and it brings that earthiness with it right back into the middle of the constructed world." (Ibid.)

17. Gormley, interview by Peat.

18. *Amazonian Field* (1991), *Field for the British Isles* (1993), and *European Field* (1993).

19. Nicholas Wroe, "Leader of the Pack," *Guardian,* June 25, 2005.

20. www.antonygormley.com.

21. Wroe, "Leader of the Pack."

Chapter 9

1. "All newly sighted subjects, indeed, have radical difficulties with appearances, finding themselves suddenly plunged into a world that, for them, may be a chaos of continually shifting, unstable, evanescent appearances. They may find themselves completely lost, at sea, in this flux of appearances, which for them is not yet securely anchored to a world of objects, a world of space. The newly sighted, who have previously depended on senses other than vision, are baffled by the very concept of 'appearance,' which, being optical, has no analogue in the other senses." (Oliver Sachs, *An Anthropologist on Mars* [New York: Knopf, 1995], 121n7)

2. "A self-directed enjoyment of nature seems to me to be something forced. More naturally, as well as more properly, we take a self-forgetful pleasure in the sheer alien pointless independent existence of animals, birds, stones, trees. 'Not how the world is, but that it is, is the mystical.' " (Iris Murdoch, *The Sovereignty of Good* [London: Routledge and Kegan Paul, 1970], 85. Quoting Ludwig Wittgenstein, *Tractatus Logico-Philosophicus,* trans. D. F. Pears and B. F. McGuinness [London: Routledge and Kegan Paul, 1963], 6.44).

3. "When you stand beneath a mature oak, or looking at a glacial lake, or at a mountain, there is a sense of being held in the presence of something that is greater in terms of time and more resilient in terms of space, rooted, present,

and the present-ness of that perception enters into your being. I think works of art aspire to this condition of present-ness and so can endow the viewer with this heightened sense of self." (Gormley, "Learning to Think")

Chapter 11

1. P. Pels, "The Spirit of Matter: On Fetish, Rarity, Fact, and Fancy," in *Border Fetishisms: Material Objects in Unstable Spaces,* ed. Patricia Spyer (New York: Routledge, 1998), 91.

2. But the Europeans, while denouncing idolatry, maintained the animist conviction that spirits speak in things—the true spirits of Christ and Saint George, Saint Michael, and Santiago.

3. "The myths tell us nothing instructive about the order of the world, the nature of reality or the origin and destiny of mankind. . . . They teach us a great deal about the societies from which they originate, they help to lay bare their inner workings and clarify the *raison d'être* of beliefs, customs and institutions . . .; most importantly, they make it possible to discover certain operational modes of the human mind. . . ." (Claude Lévi-Strauss, *The Naked Man,* trans. John and Doreen Weightman [Chicago: University of Chicago Press, 1981], 639)

4. Arthur C. Danto, *Beyond the Brillo Box* (Berkeley: University of California Press, 1992), 9; *After the End of Art* (Princeton: Princeton University Press, 1997), 30–33, 66.

5. Michael Kimmelman, *The Accidental Masterpiece* (New York: Penguin, 2005).

Chapter 12

1. George Friedrich Wilhelm Hegel, *The Philosophy of Fine Art,* trans. F. P. B. Osmaston (Bristol, England: Thoemmes Press, 1999), 38–39.

Chapter 13

1. A ten-hectare section of rain forest in Malaysia was found to contain 780 species of trees—more species than are found in all of North America.

2. Paul Shephard, *The Only World We've Got* (San Francisco: Sierra Club Books, 1996), 60.

Chapter 15

1. Henri Hubert and Marcel Mauss, *Sacrifice: Its Nature and Functions,* trans. W. D. Halls (Chicago: University of Chicago Press, 1981).

2. J. C. Heesterman, *The Broken World of Sacrifice* (Chicago: University of Chicago Press, 1993).

3. Karen McCarthy Brown, *Mama Lola: A Vodou Priestess in Brooklyn* (Berkeley and Los Angeles: University of California Press, 2001).

4. Michel Leiris, *L'Afrique fantôme* (Paris: Gallimard, 1981).

5. Thor Heyerdahl, *Aku-Aku, the Secret of Easter Island* (Chicago: Rand McNally, 1958).

Chapter 18

1. Maurice Merleau-Ponty, *The Visible and the Invisible,* trans. Alphonso Lingis (Evanston, Ill.: Northwestern University Press, 1964), 119.

Chapter 19

1. Jake (b. 1966) and Dinos (b. 1962) Chapman were born in London and attended the University of London's Art College and then the Royal College of Art.

2. The White Cube, London, 2002, The Saatchi Gallery, London, 2004, Tate-Liverpool, 2006–2007.

Chapter 20

1. Yukio Mishima, *The Temple of the Golden Pavilion* (New York: Berkley Medallion Books, 1971).

2. Martin Heidegger distinguished anxiety from fear. Anxiety is not turned to identifiable menaces and threats but instead feels a generalized danger everywhere in the world. It is, Heidegger says, as though the world is withdrawing, leaving nothing to support me, nothing to hold on to. Anxiety, he explains, is an anticipation of dying; it reveals that dying is being cast into the void, existence being hurled into nothingness. Anxiety, sensing nothingness coming to take possession of us, gives us the concept of nothingness.

Anxiety isolates me; it is my existence that I feel is cast toward nothingness, not the existence of others or of the world. This nothingness is destined uniquely for me, singles me out. The others, implanted in their environments, will continue to be there; the world will go on as before.

Anxiety can be fled or covered over. But when acknowledged, accepted, it converts, Heidegger says, into resoluteness. Feeling myself adrift in the void, nothing supporting me, nothing to hold on to, I am thrown back upon the powers that still subsist in me, the powers that are my own. No longer counting on the support of the world or of others, taking up the powers that are my own, I now resolutely act on my own, exist on my own. Thus Heidegger sees the most negative experience, the experience of nothingness itself, converting into the most positive experience, the experience that posits my existence on its own. (*Being and Time,* 279–311)

But does anxiety really identify death as nothingness? What is distinctive of anxiety, as contrasted with fear, is that anxiety has no identifiable object. It is a foreboding concerning what is indeterminate. What I am anxious about is the unknown, not something that I know to be nothingness.

We could compare the situation that arouses anxiety with our edginess in the night. What makes the night disquieting is not a perception of something dangerous, but the fact that possible dangers are not localizable in things. We no longer perceive things; their contours are dissolved by the night. But this no-thingness is not nothingness; there is left the darkness of night, which fills space, and invades, is within as much as without. The dissolution of contours and boundaries makes whatever dangers the night harbors be possibly anywhere, everywhere.

The deep uneasiness before the unknown that afflicts me does not turn into a recognition that I exist on my own and can, must, act on my own. I would instead overcome, or at least endure, anxiety by seeking support in what is known, even though it is now out of reach.

3. Pierre Clastres, *The Archeology of Violence,* trans. Jeanine Herman (New York: Semiotext(e), 1994), 152–58.

4. Clastres quite ignores this internal glory that transfigures the warrior in the moment of his deed. Clastres instead equates glory with recognition by others. (Ibid., 179)

5. Ibid., 188–89.

6. Gilles Deleuze and Félix Guattari, *Thousand Plateaus,* trans. Brian Massumi (Minneapolis: University of Minnesota Press, 1987), 418–23.

7. "These murders our laws punish so sternly, these murders we suppose the greatest outrage that can be inflicted upon Nature, not only, as you very well see, do her no hurt and can do her none, but are in some sort instrumental to her, since she is a great murderess herself. . . . The most wicked individual on earth, the most abominable, the most ferocious, the most barbarous, and the most indefatigable murderer is therefore but the spokesman of her desires, the vehicle of her will, and the surest agent of her caprices. . . . Never will too many or enough murders be committed on earth, considering the burning thirst Nature has for them." (The Marquis de Sade, *Juliette,* trans. Austryn Wainhouse [New York: Grove Press, 1968], 768–79, 772)

8. Ibid., 525.

9. Max Weber, *From Max Weber: Essays in Sociology,* trans. and edited by H. H. Gerth and C. Wright Mills (New York: Oxford University Press, 1958), 78.

Chapter 22

1. About the size of modern Los Angeles.

Chapter 23

1. Ken Beittel, *Zen and the Art of Pottery* (New York: Weatherhill, 1989), 4.

2. Marshall Sahlins, *Stone Age Economics* (New York: Aldine, 1972).

Chapter 24

1. Victor Turner, *From Ritual to Theatre: The Human Seriousness of Play* (New York: PAJ, 1982), 32.

2. Claude Lévi-Strauss *The Naked Man,* trans. John and Doreen Weightman (Chicago: University of Chicago Press, 1981), 639.

3. Clifford Geertz, *The Interpretation of Cultures* (New York: Basic Books, 1973), 115.

4. Max Weber, *The Sociology of Religion* (Boston: Beacon, 1963).

5. Edmund Leach, *The Essential Edmund Leach,* vol. 2, *Culture and Human Nature,* ed. Stephen Hugh-Hones and James Laidlaw (New Haven, Conn.: Yale University Press, 2000), 178.

6. Geertz, *Interpretation of Cultures,* 81.

7. "The study of other peoples' cultures . . . involves discovering who they think they are, what they think they are doing, and to what end they think they are doing it." (Clifford Geertz, *Available Light* [Princeton, N.J.: Princeton University Press, 2000], 16)

8. Ibid., 362–63.

9. Ibid., 134–35.

10. "Symbols . . . are tangible formulations of notions, abstractions from experience fixed in perceptible forms, concrete embodiments of ideas, attitudes, judgments, longings, or beliefs." (Ibid., 91)

11. Ibid., 449.

12. Ibid., 177, 179.

13. Donald Cordry, *Mexican Masks* (Austin and London: University of Texas Press, 1980), 23–31. "A masking tradition can survive even when its original symbolism and relation to social context have been completely lost. . . . Should we then conclude that such masked celebrations are empty charades or romantic anachronisms? Quite the contrary; I think that they demonstrate two things: the power of the masked disguise in and of itself, and its vital importance to the social community even in the absence of specific meaning." (J. C. Crocker, "Ceremonial Masks," in Victor Turner, ed., *Celebration: Studies in Festivity and Ritual* [Washington, D.C.: Smithsonian Institution Press, 1982], 80)

14. Barbara Babcock, "Too Many, Too Few: Ritual Modes of Signification," *Semiotica* 23 (January 1978), 296.

15. Don't we sense that rituals in public life replace meaningful public participation and meaningful policies, that the daily rituals in our personal or professional lives take the place of meaning in our days and lives that is lacking?

16. Geertz, *Interpretation of Cultures,* 118–19, 181.

17. Emmanuel Levinas has incisively distinguished consciousness subjected to rhythm from the active, projective movement of intentionality. (Emmanuel Levinas, "Reality and Its Shadow," in Emmanuel Levinas, *Collected Philosophical Papers,* trans. Alphonso Lingis [The Hague: Martinus Nijhoff, 1987], 4–5)

18. He adds "The recent work on the neurobiology of the brain . . . shows, among other things, how the 'driving techniques of ritual (including sonic driving, for example, percussion instruments) facilitate right hemisphere domi-

nance, resulting in gestalt, timeless, nonverbal experiences, differentiated and unique when compared with the left hemisphere functioning or hemisphere alternation.'" (Victor Turner in Victor Turner, ed, *The Anthropology of Experience* [Urbana: University of Illinois Press, 1986], 43).

19. Phenomenology has to distinguish this consciousness enthralled by the spectacular in heightened intensity from the consciousness that actively grasps objects in forming concepts.

Chapter 25

1. Paula Brown, *Highland Peoples of New Guinea* (Cambridge: Cambridge University Press, 1978), 156.

2. Friedrich Nietzsche, *The Gay Science,* trans. Walter Kaufmann (New York: Vintage, 1974), § 337.

3. Kenneth Read recounts the ritual he attended among the Gahuku in the New Guinea Highlands:

"The house was packed to its capacity, but in the blackness I was unable to discover so much as a single feature of the man who sat beside me. Almost immediately, enveloped in disembodied voices, I felt the first stirrings of a curious panic, a fear that if I relaxed my objectivity for so much as a moment I would lose my identity. At the same time the possibility that this could happen seemed immensely attractive. The air was thick with pungent odors, with the smell of unwashed bodies and stranger aromatic overtones that pricked my nostrils and my eyes. But it was the singing, reverberating in the confined space and pounding incessantly against my ears that rose to cloud my mind with the fumes of a collective motion almost too powerful for my independent will. Momentarily the night vanished, and my purpose, even the circumstances of my presence in the village were no longer important. I stood poised at a threshold promising a release from the doubts and anxieties that separate us from one another, offering, if one took the step demanded, a surety, a comforting acceptance such as those who share an ultimate commitment may experience. Even though the words were unintelligible, the massed voices were like a hand held toward me, a proffered embrace.

"It was this thought, or, rather, this intuition, for it was hardly a conscious reaction, that held in check my feeling of suffocation. The songs followed one another without a perceptible break, a single shrill and keening voice lifting now and then to point the way to a new set. As the others joined in strongly, I felt close to the very things that eluded me in my day-to-day investigations, brought into physical confrontation with the intangible realm of hopes and shared ideas for which words and actions, though they are all we have, are quite inadequate expressions. In analytic language, the situation could be accommodated under the rubric of a rite of separation—an event by which a young girl in her father's house, surrounded by her kinsmen, was brought to the morning of the day on which she must assume a new status and be transferred to her husband's people, but its quality could not be conveyed in any professional terms. While the voices swelled inside the house, mounting to a climax, the barriers of my alien life dissolved. The sound engulfed me, bearing me with it beyond the house and into the empty spaces of the revolving universe. Thus sustained, I was one of the innumerable companies of men who, back to the shrouded entrance of the human race, have sat at night by fires and filled the forest clearings and the wilderness with recitals of their own uniqueness." (*The High Valley* [New York: Scribner's, 1965], 251–52)

4. Ornithologists have experimentally verified that female peafowl, sage grouse, and birds-of-paradise select for their sexual favors males that display with the most elaborate dances the most spectacular plumage, even though their ostentatious colors and entranced dances make them easy prey for predators, and the males of these species contribute nothing to the nest building and guarding and nurturing tasks that ensure the reproduction of the species.

5. Friedrich Nietzsche, *Thus Spoke Zarathustra,* trans. Walter Kaufmann in *The Portable Nietzsche* (New York: Viking, 1968), book 3, "The Convalescent," 2.

Photographs

Photographs are by the author, unless otherwise noted.

About the Author

Alphonso Lingis is a professor emeritus of philosophy at Penn State University and the author of numerous books, including *Abuses, The Imperative, Dangerous Emotions, Trust, Body Transformations, The First Person Singular,* and most recently, *Contact.* He is also the preeminent English translator of Maurice Merleau-Ponty and Emmanuel Levinas. His translations include Merleau-Ponty's *The Visible and the Invisible* and Pierre Klossowski's *Sade My Neighbor,* both published by Northwestern University Press.